Marine Transportation and International Trade 2018

Transportation Research Record: Journal of the Transportation Research Board

Volume 2672, Issue 11

ISBN: 978-1-5108-7732-0

Printed from e-media with permission by:

Curran Associates, Inc.
57 Morehouse Lane
Red Hook, NY 12571

Some format issues inherent in the e-media version may also appear in this print version.

Published by Sage Publications USA

ISBN (Print) 978-1-5108-7732-0
ISBN 2018 Printed Set (All Issues) 978-1-5108-7735-1

Printed by Curran Associates, Inc. (2018)

Additional copies of this publication are available from:

Curran Associates, Inc.
57 Morehouse Lane
Red Hook, NY 12571 USA
Phone: 845-758-0400
Fax: 845-758-2633
Email: curran@proceedings.com
Web: www.proceedings.com

Volume 2672 Number 11 December 2018

TRR
JOURNAL OF THE TRANSPORTATION RESEARCH BOARD

Contents

Marine Transportation and International Trade

Articles

Article

TRR
JOURNAL OF THE TRANSPORTATION RESEARCH BOARD

Transportation Research Record
2018, Vol. 2672(11) 1–9

DOI: 10.1177/0361198118756622
journals.sagepub.com/home/trr
SAGE

The Dynamics of Dry Bulk Shipping Market Under the Shipping Cycle Perspective: Market Relationships and Volatility

Yijie Wu[1], Jingbo Yin[1], and Pan Sheng[1]

Abstract

The shipping industry plays an essential role in world trade. For shipping companies, having an accurate view of the markets and grasp of the interactions between the freight market, second-hand ship market, and the newbuild ship markets is essential. The shipping market cycles are divided into four periods (trough, recovery, peak, and recession) based upon shipping cycle theory. The current shipping markets have been stuck in the trough period since the financial crisis in 2008. This paper investigates the recovery period and causality relationship between the freight rate, second-hand price, and new-build ship price, in the dry bulk shipping market and applies the Granger causality test at each stage of the cycle based on quantitative analysis. The results show that the recovery period and causality relationship can be identified only during the trough and peak periods. When comparing the results for the trough period before and after the financial crisis, we find similarities between the two periods, leading us to conclude that the shipping cycle rules still apply.

As an important branch of the shipping markets, the dry bulk shipping market, mainly concerns industrial materials such as iron ore, coal, and cereals. In the dry bulk shipping market, there are many small-scale ship owners, no market barriers for market participants, and all transactions tend to be transparent. The cyclical fluctuations within the dry bulk shipping market have been divided into three cycles: the long shipping cycle, the short business cycle, and the seasonal cycle. It is important to understand how they fit into the economics of the shipping markets, as oil prices, war, and technology innovation all impact shipping market cycles (*1*).

The three cycles can be divided into trough, recovery, peak, and recession periods (*2*). The level of volatility, the relationships that exist between the shipping markets, freight rates and ship prices all fluctuate for each period of the shipping cycle, which requires different investment strategies should be taken for each period by the individual shipping companies. To adopt the most suitable investment strategy in rapidly changing markets, companies not only have to be familiar with their own market and able to react to volatility in the markets, so shipping cycle theory is an important tool in making predictions regarding the shipping markets.

The financial crisis in 2008 caused the maritime environment to change considerably, impacted shipping cycles. The shipping markets have been stuck in a trough period since this time, with little indication that they will enter a recovery period in the near future. There is a view that the shipping cycles no longer exist, and also a suggestion that the shipping cycles do exist, but the duration of each period has changed over this time of recession. Dinwoodie and Landamore collected several specialists' opinions of the shipping cycle in 2015 and most of them agreed that the shipping market will recover before 2050 (*3*).

This paper tries to investigate the internal relationships between the shipping markets before and after the financial crisis of 2008, for each period of the shipping cycle. Fluctuation in one market is likely to be an indication of change in another market, and these changes can have effects across the different markets. For example, fluctuation in the freight market will reflect on the second-hand ship market and the new-build ship market; the ups and downs of second-hand ship price will have an influence on the new-build ship

[1]Department of International Shipping, School of Naval Architecture, Ocean and Civil Engineering, Shanghai Jiaotong University, Minhang, Shanghai, China

Corresponding Author:
Address correspondence to Jingbo Yin: jingboyin@sjtu.edu.cn

price; and, at the same time, fluctuation in the ship market may have an impact on the freight market. It is important to study the volatility within the markets and the relationships that exist among them.

This paper uses quantitative analysis to investigate the recovery period and causality relationship that exists among the freight rate, second-hand price, and new-build ship price in the dry bulk shipping market. Owing to the differences in shipping cycle patterns before and after the financial crisis, this paper conducts a recovery period test and Granger causality test at each stage of the shipping cycle (trough, recovery, peak, and recession) for the dry bulk shipping market.

Literature Review

Freight markets and ship markets are studied to better understand the market discipline and to investigate the underlying relationships in the different markets, both at a theoretical and an empirical level.

Tinbergen's study analyzed the relationships between shipping markets and studied the cyclicality of the shipbuilding market and pointed out that the new-build ship market is dependent on the freight rate, which is, in turn, dependent on the shipping capacity provided by the new-build market (*4*). The time delay between the demand for shipping capacity and the supply of shipping capacity leads to an endogenous market cycle. Koopmans (*5*) used the cobweb theorem and suggested that the new-build market is influenced by expectations concerning the degree of equilibrium between shipping capacity and transportation demand. Later, by taking the time delay into consideration and based on the cobweb theorem, Tinbergen used a supply-and-demand approach to analyze the new-build market and reached the conclusion that low shipping tonnage leads to high freight rates.

A variety of models have been used to analyze the relationships between markets and to predict the future trends of markets. Beenstock described a theoretical model in which ship markets and freight markets are interdependent and the ship is a capital asset (*6*). Beenstock and Vergottis took freight, the second-hand ship price, and the new-build ship price into consideration to predict the world market for dry cargo (*7*). They found that an increase in the freight rate has little influence on ship price and there is no time delay between them. However, shipping market data reveals that the ship price reaches its peak several months after the freight rate reaches its peak. The cyclical nature of the shipbuilding market lags behind the cyclicality of the freight markets. Beenstock and Vergottis reviewed the new-build ship prices and market and their relationship with the second-hand ship market (*8*). They demonstrated that shipbuilding costs have no influence on the new-build ship price, in contrast to the economic theory that shipbuilding costs have a large influence on new-build ship price. Volk combined the capital asset pricing model (CAPM) with the cost-based model to explain the relationship between the shipping and freight markets (*9*).

Time series methods have been used to study the relationships among different markets. Kavussanos employed a vector auto-regression (VAR) model, an auto-regressive conditional heteroskedasticity (ARCH) model, and an auto-regressive and moving average mode (AR(I)MA) model to estimate second-hand ship prices to examine the dynamics of conditional volatilities for second-hand ships in the world dry bulk market (*10*). Using ARCH models and monthly data to model and compare volatility estimates between different-sized vessels, Kavussanos found that, in general, the prices of small vessels are less volatile than those of larger ones and the nature of the volatilities varies across sizes.

Hale and Vanages (*11*), Glen (*12*), and Glen and Martin (*13*) found recovery period to be an excellent method for research when examining the relationships between ship markets and shipping markets. Using the recovery period methodology, Hale and Vanages established that second-hand ship prices for various ship sizes are stationary in first differences, thus allowing the search for a long-term recovery period relationship among them. To investigate the performance of trading strategies based on a combination of technical trading rules and fundamental analysis in the sale and purchase market, Alizadeh and Nomikos (*14*) established a long-term relationship between price and income using the recovery period methodology and stationary bootstrap approach. The results indicated that trading strategies based on the earnings–price ratios are superior to buy-and-hold strategies in the tanker market.

Research after the financial crisis of 2008 still focused on relationships and volatility of the shipping market. Duru (*15*) choose the fuzzy time series method to study the volatility of the shipping market. Using the GARCH model, Dai and Hu (*16*) built a model of the relationship between the price and the freight rate. Xu (*17*) adopted the recovery period methodology to study the interaction between the freight market and the newbuilding ship market, and put forward questions regarding the causality between the freight rate and shipbuilding price, the determinants of shipbuilding activities, and the impact of the change of fleet size on freight volatility. By applying the unit root test, recovery period test, and Granger causality test, Xu pointed out that the fluctuations in the freight rate resulted in fluctuations in the new-build ship price, and that there was generally a 3-month time lag. In contrast to earlier research mentioned previously, Xu found that the new-build ship price cannot influence the freight rate.

Consensus among most scholars today is that studies of the shipping market should be divided into two periods: the period before the financial crisis of 2008, and the period after the financial crisis. It is considered that the relationships among the freight market, new-build ship market, and second-hand ship market should each be calculated independently. The existing research achievements based on the data before 2008 may not suitable for the new period and there has been little comparison or research of the market between two periods.

Methodology

The relationships between the shipping markets are changing regarding shipping cycles. This paper focuses on the influences among the markets and employs time series methods, including Johansen's recovery period test and Granger causality test, to investigate the interrelationships between the shipping markets at each stage of the shipping cycle. By applying the recovery period test, influences between the markets exit can be confirmed and, by applying the causality test, further time relationships among the markets may be found.

Johansen Test

If two or more series are individually integrated (in the time series sense) but some linear combination of them has a lower order of integration, then the series are said to be co-integrated and the linear combination is called the co-integrating equation. The recovery period equation allows us to test whether the variables we have chosen to include in our model exhibit such relationships and, consequently, can be used in the model.

If the recovery period relationship exists between two or more variables, we can conclude a long-term stable relationship. A change in one of the variables will influence another. In contrast, if the recovery period relationship does not exist, then there is no long-term stable relationship between them.

There are two common ways to test for recovery period: the Engle–Granger (EG) two-step method and the Johansen test. Johansen's test for recovery period uses the VAR approach because it is based on a multi-variant framework, which allows us to choose the number of lags and take into account both endogeneity and simultaneity problems (*12*). In this paper, the Johansen test was chosen to Test for Recovery Period. The equation of the VAR model is

$$Y_t = B_1 Y_{t-1} + B_2 Y_{t-2} + \ldots + B_p Y_{t-p} + U_t \qquad (1)$$

where:

$Y_t = \left(Y_{1t}\, Y_{2t} \cdots Y_{mt}\right)'$ is the m-dimension stochastic vector;

$B_i = \begin{pmatrix} B_{i11} & B_{i12} & \cdots & B_{i1m} \\ B_{i21} & B_{i22} & \cdots & B_{i2m} \\ \vdots & \vdots & \ddots & \vdots \\ B_{im1} & B_{im2} & \cdots & B_{imm} \end{pmatrix}$ (i =1,2, …, p), is the parameter matrix;

$U_t \sim \mathrm{IID}(0, \Sigma)$ is the error term.

The trace test (λ_{trace}) and the maximum Eigenvalue test ($\lambda_{\max}$) can be employed to test the number of co-integrating equations.

The null hypothesis of the λ_{trace} test is $H_0 : r < m \leftrightarrow H_1 : r = m$ (r is the number of characteristic roots, m is the number of co-integrating equations in the hypothesis), and the null hypothesis of the $\lambda_{\max}$ test is $H_0 : r = q,\ q = 1, 2, \cdots, m \leftrightarrow H_1 : r \le q + 1$ (q is the number of co-integrating equations in the hypothesis).:

$$\lambda_{\text{trace}} = -T * \sum_{i=r+1}^{m} \ln\left(1 - \lambda_i\right) \qquad (2)$$

$$\lambda_{\max} = -T \ln\left(1 - \lambda_{r+1}\right) \qquad (3)$$

where:

λ_i = the estimated value of the eigenvalue of Y_t; and

T = the sample size.

These tests were conducted sequentially according to the number of r characteristic roots. Assuming the value of r is zero, and the number of co-integrating equations is zero. If the test results reject the null hypothesis H_0, it represents one recovery period relationship, continue testing. If the results still reject H_0, it represents m recovery period relationships. Testing was repeated until the results no longer reject H_0. The more recovery period vectors in the model, the more stable the system.

Granger Causality Test

The Granger causality test is used to judge the causality of different variables. For the two variables X and Y, if a change in X can explain a future change in Y, then X is the Granger cause of Y. By using the Granger causality test, we can establish if one variable will influence fluctuation in the other variables. The equations for the Granger causality regression are

$$y_t = \sum_{i-1}^{p} \alpha_i x_{t-i} + \sum_{j-1}^{p} \beta_j y_{t-j} + u_{1t} \qquad (4)$$

$$x_t = \sum_{i-1}^{p} \lambda_i x_{t-i} + \sum_{j-1}^{p} \delta_j y_{t-j} + u_{2t} \qquad (5)$$

where:

u_{1t} and u_{2t} = two uncorrelated white noise time series;

p = the largest lag duration of the model; and

$\alpha_i, \beta_j, \lambda_i, \delta_j$ are the coefficients of the model.

If the lag coefficients α_i are significant while the lag coefficients δ_j are not significant, then X is the Granger causality of Y. That means variable X has an impact on the fluctuation of variable Y with a fixed time lag.

Data Description

The data used in this paper come from the Clarkson Shipping Intelligence Network (SIN). The Baltic Dry Index (BDI) is used to represent fluctuations in the freight market [for the period before October 1999, the Baltic Freight Index (BFI) was used instead of the BDI]. The Newbuilding Bulkcarrier Price index (NBP) is used to represent fluctuations in the newbuilding ship market. The Bulk Carrier Second-Hand Price index (SHP) for ships 5 years old or older is used to

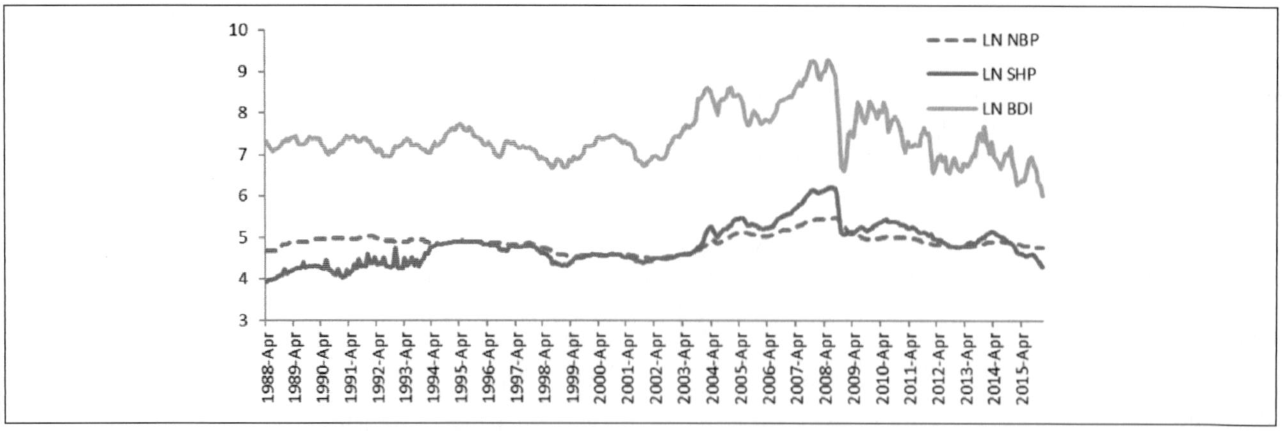

Figure 1. The fluctuation of BDI, NBP, and SHP.

represent fluctuations in the second-hand ship market. All the variables are monthly data, in U.S. dollars per ton. Fluctuations in these three indices for 1988–2016 are shown in Figure 1. Co-movement in these series indicates that some inner relationships exist among the three indices (Figure 1).

The dry bulk shipping market shows different regulations before and after the financial crisis in 2008, and the two time periods should be analyzed, respectively. The BDI started to rise abnormally in July 2004 and stayed at a high level until the financial crisis in 2008. The dry bulk shipping market shows significant differences over this period and is not considered to represent standard patterns and so is not included in this paper. Each stage prior to 1994 usually lasted 1–2 years, which is not significant as a test period, and also not included in this paper. Accordingly, the dry bulk shipping market is divided into four periods prior to July 2004 and the period after 2008 belongs to another trough period (Figure 2 and Table 1).

Empirical Analysis

For each period of the shipping cycle, the relationships between the three markets were analyzed using recovery period theory. As the procedure for the recovery period test and the Granger causality test is the same for each period, we only show details of these tests for the trough period after 2008 as an example. At the end of this section, the results for each period are presented with comparisons made.

Stationary Test

According to recovery period theory, only the data that have smooth stationarity or show stationarity after first differencing are eligible for further analysis. The results of the unit root test for the trough period after 2008 are shown in Table 2.

According to the results of the unit root test, although the original data show nonstationarity, the first-order differences are stationary. Hence, the dynamics and intercorrelations among the three variables (BDI, SHP, and NBP) can be further analyzed using the recovery period test.

Recovery Period Test

The recovery period test can determine whether the recovery period relationship exists among the variables, as well as the number of recovery period equations. For this paper, recovery period test, five different criteria used to determine the time lag (Table 3).

The results suggest that the most suitable time lag is 2 years, while only one criterion suggests that the time lag should be 4 years (see values in bold in Table 3). Therefore, we chose 2 as the time lag among the variables for the trough period after the financial crisis in 2008. The results of the Johansen tests, namely the trace test and maximum Eigenvalue test, are reported in Table 4.

According to the results of both Johansen recovery period tests, we conclude that the assumption that no recovery period exists among the three markets is incorrect, but the other assumption, that at most one recovery period exists, is found to be true. The results reflect that only one recovery period relationship exists among the freight market, newbuilding ship market, and second-hand ship market for the trough period after 2008. Based on the results, we build a VAR model for the trough period after 2008:

$$\begin{aligned} \mathrm{SHP}_t = {} & 0.000881\mathrm{BDI}_{t-1} + 0.000837\mathrm{BDI}_{t-2} \\ & + 0.350325\mathrm{NBP}_{t-1} - 0.372990\mathrm{NBP}_{t-2} \\ & + 1.121544\mathrm{SHP}_{t-1} - 0.131480\mathrm{SHP}_{t-2} \\ & + 2.073543 \end{aligned} \quad (6)$$

$$\begin{aligned} \mathrm{NBP}_t = {} & -0.000041\mathrm{BDI}_{t-1} + 0.000090\mathrm{BDI}_{t-2} \\ & + 1.103965\mathrm{NBP}_{t-1} - 0.129890\mathrm{NBP}_{t-2} \\ & + 0.044197\mathrm{SHP}_{t-1} - 0.038222\mathrm{SHP}_{t-2} \\ & + 2.362952 \end{aligned} \quad (7)$$

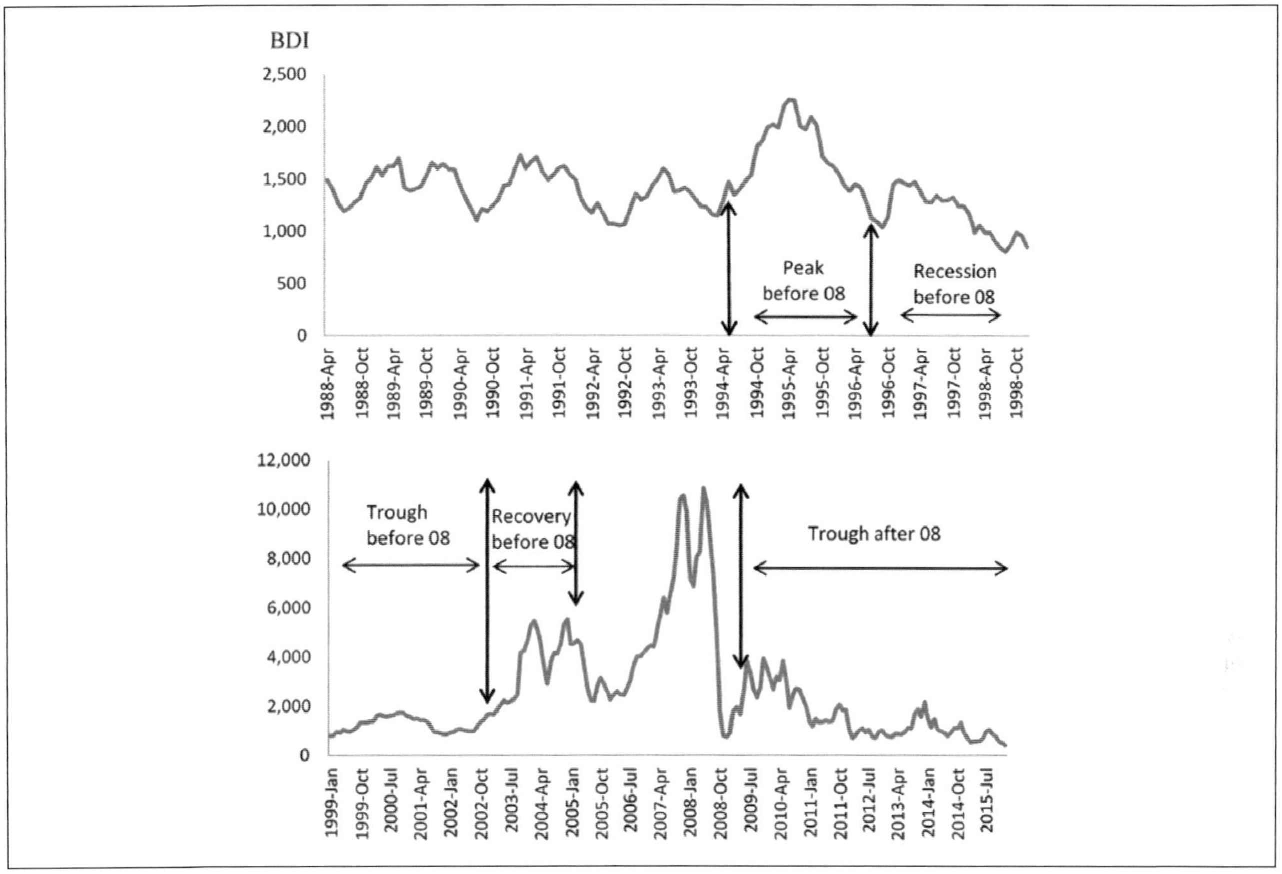

Figure 2. Partition of time series through shipping cycle.

Table 1. Partition of Time Series through Shipping Cycle

Period	Data time span	Mean BDI (BFI)	Mean SHP	Mean NBP
Peak before 2008	09/1994–03/1997	1654	125	131
Recession before 2008	04/1997–08/1998	1139	110	122
Trough before 2008	09/1998–05/2002	1230	90	96
Recovery before 2008	06/2002–06/2004	2426	107	102
Trough after 2008	01/2009–01/2016	1076	129	138

$$\begin{aligned}\mathrm{BDI}_t &= 1.549863\mathrm{BDI}_{t-1} - 0.621099\mathrm{BDI}_{t-2} \\ &- 13.09448\mathrm{NBP}_{t-1} + 11.34738\mathrm{NBP}_{t-2} \\ &- 1.918225\mathrm{SHP}_{t-1} + 2.967804\mathrm{SHP}_{t-2} \\ &+ 152.2920\end{aligned} \tag{8}$$

Granger Causality Test

The existence of a recovery period suggests long-term correlation exists among the variables. The Granger causality test was conducted to analyze the causality among the variables for the trough period after 2008. The time lag for the model is still 2. The results of the Granger causality test for

Table 2. The Result of the Unit Root Test in the Trough Period after 2008

Variable	ADF *t*-statistics	*p*-value	Test result
Original sequence			
BDI	−2.011099	0.2817	Nonstationary
SHP	0.238937	0.9735	Nonstationary
NBP	−2.389814	0.1478	Nonstationary
First-order difference sequence			
ΔBDI	−8.922952	0.0000	Stationary
ΔSHP	−6.767149	0.0000	Stationary
ΔNBP	−3.098960	0.0306	Stationary

Note: **Δ** = the data after first-order differencing; ADF = augmented Dickey–Fuller test.
p-value is acceptable under 5% level of significance.

Table 3. The Time Lag of Variables in the Trough Period after 2008

Lag	LR	FPE	AIC	SC	HQ
0	NA	8.42×10^{-5}	−0.868356	−0.805617	−0.842861
1	1247.339	1.06×10^{-8}	−9.845361	−9.594403	−9.743380
2	51.04341	8.26×10^{-9}*	10.09862*	−9.659445*	−9.920156*
3	10.64095	8.66×10^{-9}	−10.05219	−9.424796	−9.797239
4	19.90669*	8.42×10^{-9}	−10.08005	−9.264438	−9.748615

Note: * = the recommended time lag under this criterion; LR = likelihood ratio test statistic (5% level); FPE = final prediction error; AIC = Akaike information criterion; SC = Schwarz information criterion; HQ = Hannan–Quinn information criterion.

Table 4. The Result of the Johansen Recovery Period Test in Trough Period after 2008

Assumed equation number	Eigenvalue	Trace statistic	5% critical value	*p*-value
Johansen Trace Statistic Test				
None *	0.095340	35.26359	29.79707	0.0106
At most 1	0.033010	9.012307	15.49471	0.3642
At most 2	0.000830	0.217676	3.841466	0.6408
Johansen Maximal Eigenvalue Test				
None *	0.095340	26.25128	21.13162	0.0087
At most 1	0.033010	8.794631	14.26460	0.3036
At most 2	0.000830	0.217676	3.841466	0.6408

Note: * = reject null hypothesis under 5% level of significance.

Table 5. The Result of the Granger Causality Test in the Trough Period after 2008

Original hypothesis	*F*-statistic	*p*-value	Accept/reject
SHP is not the Granger reason of BDI*	2.99746	0.0120	Reject
BDI is not the Granger reason of SHP*	4.95274	0.0002	Reject
NBP is not the Granger reason of BDI	1.32973	0.2520	Accept
BDI is not the Granger reason of NBP	2.04582	0.0728	Accept
NBP is not the Granger reason of SHP*	3.06489	0.0105	Reject
SHP is not the Granger reason of NBP*	3.30496	0.0066	Reject

Note: * = reject null hypothesis under 5% level of significance.

this period are shown in Table 5. For the trough period after 2008, SHP is the Granger reason for BDI and NBP. BDI and NBP are the Granger reasons for SHP, while there is no causality between BDI and NBP.

The result shows that long-run correlation and influence can be found between the second-hand market and the other two markets, whereas there is no notable causality between the new-build market and the freight market. The results differ from most other similar studies, as other studies usually show that BDI is the Granger cause of NBP and SHP. One reason for this could be that other researchers undertake the Granger causality test using all the time series data for the shipping markets, whereas our results are based on the trough period after 2008 only. A deeper reason for the nonexistence of a long-run correlation between BDI and NBP may be the time lag of the new-build ship to the freight market. This may lead investors or shipping companies to consider other factors such as the influence of ship maximization in the 2-year time lag in addition to the influence of current market when considering a new-build ship order. The number of new-build ship orders also impacts on the new-build ship price. Investors judgments of future markets may also influence the new-build decision, as these opinions can be totally different especially in a trough period. The influence of these factors can be concealed by the brisk market in the peak period while the influence can be revealed for the trough period, and we would not consider the causality between BDI and NBP to be significant in the trough period.

Table 6. Summary of Results for Each Period

Time period	Stationary Test (First Differencing)	Recovery Period Test	Granger Causality Test
Trough before 2008	Stationary	Solo Recovery Period relationship	Exist causality (SHP→NBP; SHP→BDI)
Recovery before 2008	Stationary	Multiple Recovery Period relationship	Multiple causality
Peak before 2008	Stationary	Solo Recovery Period relationship	Exist causality (BDI→SHP)
Recession before 2008	Stationary	No Recovery Period relationship	No causality
Trough after 2008	Stationary	Solo Recovery Period relationship	Exist causality (SHP ↔ NBP; SHP ↔ BDI)

Note: SHP→BDI = the second-hand market is the Granger causer of the freight market.

Comparison among Different Periods

All the results for the stationarity test, recovery period test, and causality test for each period are shown in Table 6.

The results show that the relationships among the freight market, new-build market, and second-hand market for each period differ slightly, which means shipping companies and ship owners are likely to adopt different investment strategies for each shipping cycle period. For example, in a trough period, companies can make decisions after observing the second-hand ship market as it will have an influence on the other two markets, as the results of the causality test for this period show.

Trough Period. First, we focus on the trough period as this is the current period. Some aspects of the results for the trough period both before and after the financial crisis of 2008 show similar results. The recovery period test results show that the variables have one recovery period relationship. For the result of Granger causality test, only the second-hand prices show a causal relationship to other two markets for these two trough periods (Figure 3).

From the Granger causality test, we find that the relationships among the three markets are similar for the two trough periods; in both trough periods, the second-hand ship market influences the freight market and new-build ship market. There is no causality relationship between the freight market and new-build ship market, yet some characteristics of the trough period of the shipping market were maintained after the financial crisis, despite considerable market changes. For the trough period before 2008, the causality relationships between the second-hand ship market and the other two markets are unidirectional, which means a change of the second-hand ship price directly influences the other two markets, but the converse is not true. For the trough period after 2008, the causality relationships between the second-hand ship market and other two markets are bidirectional, which means the second-hand ship market and other two markets influence each other.

The trough stage after 2008 has lasted more than 8 years, which is almost twice as long as a general trough period prior to the financial crisis. In a longer trough period, a second-hand ship becomes a limited resource within a depressed new-build ship market and brisk demolition market. As the pool of second-hand vessels dries up, the market may be driven to the more expensive and drawn out prospect of new-builds to meet demand. In this case, the new-build ship market and the freight market became the decisive factors in the shipping market, instead of the second-hand ship market, with the freight market and new-build market becoming the Granger causes of the second-hand market, and showed the bidirectional relationships between the markets in a longer trough stage, as is the current situation.

Recovery Period. The recovery period is different from other periods due to the level of recovery period among the three market cycles. In addition, the Granger causality test shows that many circumstances of causality exist. We suggest that there are two main causes for this: the recovery period generally lasts a short time, mostly between 1 and 2 years. In this case, we had data for a 12–24-month period for the recovery period tests, which may lead to multiple results; and difficulties in defining the beginning and end of recovery period may lead to multiple results.

Peak Period. For the peak period, the recovery period test, and the Granger causality test show that there only exists one recovery period relationship and one causality among the three markets. The freight market has an impact on the fluctuation of the second-hand ship market whereas other Granger causalities do not exist during the peak stage (Figure 3).

Compared with the trough period before 2008, the relationship between the freight market and the second-hand ship market is just contrary. Another difference is that the second-hand market will influence the new-build market in the trough period whereas this causality does not exist in the peak period. The primary cause for this is the time lag in ordering a new ship. The 2-year time delay may be acceptable in the trough period, but a second-hand ship is preferable in a brisk market as the shipping companies can expand as soon as possible in such a market. For this reason, BDI

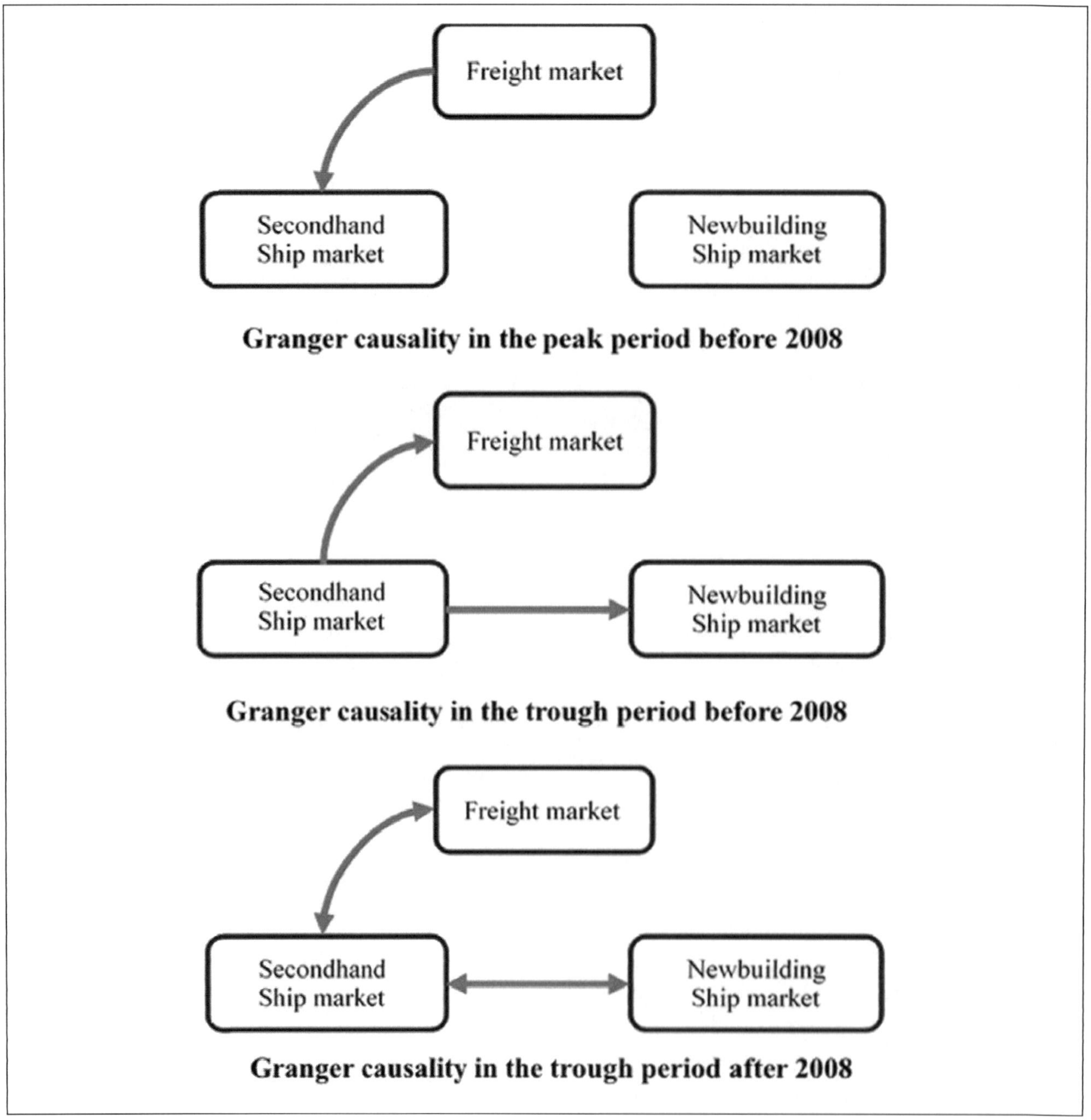

Figure 3. The results of Granger causality test in three periods.

will influence SHP whereas no other direct relationships exist among the three markets in the peak period.

Recession Period. The recession period lasted roughly 6–12 months, which does not suffice as a test period. Owing to the lack of data, this period was not taken into consideration in this study.

Conclusion

This paper has focused on the correlations among the different shipping markets and the shipping cycles before and after the financial crisis in 2008. The difference in market relations and volatility for each period can help us understand the market changes.

Time series methods, including the unit root test, recovery period test, and Granger causality test, were used to analyze the correlations among the different shipping markets. The results concluded that the relationships between the freight market, new-build market, and second-hand market for each period are different. First, the recovery period test suggests that only one recovery period relationship is found during the trough and peak periods for the three markets with a 2-month time lag, yet for other periods, the recovery period relationship is not unique in the recovery period and does not exist in the recession period. Second, significant causality only exists for the peak and trough periods, both before and after the crisis. The Granger causality test illustrates that the relationship between the freight market and the second-hand ship market is reversed in the trough and peak periods. As the correlations among these three markets are different for different cycles, shipping companies and ship owners should adjust their approaches to investment, taking into consideration the different variables.

The Granger causality test results for the respective trough periods are strikingly similar. In both periods, the second-hand ship price influences the other two markets, and there is no causality relation between the freight market and new-build ship market. This indicates the second-hand ship price is related to the market volatility while the new-build ship price may also be influenced by other factors like the investors' individual judgments. Comparing the results of the two trough periods, we find similarities, which suggests a connection between the trough period before and after the crisis. This finding leads us to the conclusion that the shipping cycle rules can still be applied to the shipping markets today.

According to the results, companies can make decisions after observing the second-hand ship market in the trough stage as it will have an influence on the other two markets. The existence of bidirectional relationships among current shipping markets also reminds investors to pay attention to the new-build and freight markets. Further research on this topic may focus on the specific investment strategies in current trough stage of the shipping cycle, where economic models in decision-making can be applied during different shipping cycle stages. Other factors such as investment risk and the rate of return can also be utilized to find the most suitable investment strategic in the future research.

References

1. Merika, A., and S. Theodoropoulou. The Relationship between Business Cycles and Capital Structure Choice: The Case of the International Shipping Industry. *The Journal of Economic Asymmetries*, Vol. 12, No. 2, 2015, pp. 92–99.
2. Lai, K. K., K. Lam, and W. K. Chan. Shipping Container Logistics and Allocation. *Journal of the Operational Research Society*, Vol. 46, No. 6, 1995, pp. 687–697.
3. Dinwoodie, J., and M. Landamore. Dry Bulk Shipping Flows to 2050: Delphi Perceptions of Early Career Specialists. *Technological Forecasting and Social Change*, Vol. 88, 2014, pp. 64–75.
4. Tinbergen, J. Ein Schiffbauzyklus. *Articles*, Vol. 34, 1931, pp. 152–164.
5. Koopmans, T. Tanker Freight Rates and Tankship Building: An Analysis of Cyclical Fluctuations. *De erven F. Bohn nv*, No. 27, 1939.
6. Beenstock, M. A Theory of Ship Prices. *Maritime Policy and Management*, Vol. 12, No. 3, 1985, pp. 215–225.
7. Beenstock, M., and A. Vergottis. An Econometric Model of the World Market for Dry Cargo Freight and Shipping. *Applied Economics*, Vol. 21, No. 3, 1989, pp. 339–356.
8. Beenstock, M., and A. Vergottis. *Econometric Modelling of World Shipping*. Springer Science & Business Media, 1993.
9. Volk, B. *The Shipbuilding Cycle: A Phenomenon Explained.* Institute of Shipping Economics and Logistics, 1994.
10. Kavussanos, M. G. The Dynamics of Time-Varying Volatilities in Different Size Second-Hand Ship Prices of the Dry-Cargo Sector. *Applied Economics*, Vol. 29, No. 4, 1997, pp. 433–443.
11. Hale, C., and A. Vanages. The Market for Second-Hand Ships: Some Results on Efficiency using Recovery Period. *Maritime Policy and Management*, Vol. 19, No. 1, 1992, pp. 31–39.
12. Glen, D. The Market for Second-Hand Ships: Further Results on Efficiency using Recovery Period Analysis. *Maritime Policy and Management*, Vol. 24, No. 3, 1997, pp. 245–260.
13. Glen, D., and B. Martin. Conditional Modelling of Tanker Market Risk using Route Specific Freight Rates. *Maritime Policy and Management*, Vol. 25, No. 2, 1998, pp. 117–128.
14. Alizadeh, A. H., and N. K. Nomikos. Investment Timing and Trading Strategies in the Sale and Purchase Market for Ships. *Transportation Research Part B: Methodological*, Vol. 41, No. 1, 2007, pp. 126–143.
15. Duru, O. A Fuzzy Integrated Logical Forecasting Model for Dry Bulk Shipping Index Forecasting: An Improved Fuzzy Time Series Approach. *Expert Systems with Applications*, Vol. 37, No. 7, 2010, pp. 5372–5380.
16. Dai, L., and H. Hu. An Empirical Analysis of Freight Rate and Vessel Price Volatility Transmission in Global Dry Bulk Shipping Market. *Journal of Traffic and Transportation Engineering (English Edition)*, Vol. 2, No. 5, 2015, pp. 353–361.
17. Xu, J. J. *The Dynamics of Shipbuilding in Shipping: An Econometric Analysis*. The Hong Kong Polytechnic University, 2010.

The Standing Committee on International Trade and Transportation (AT020) peer-reviewed this paper (18-05845).

Article

TRR
JOURNAL OF THE TRANSPORTATION RESEARCH BOARD

Transportation Research Record
2018, Vol. 2672(11) 10–18

DOI: 10.1177/0361198118758295
journals.sagepub.com/home/trr
SAGE

Assessment of Canada's Transportation System under the Comprehensive Economic and Trade Agreement

Mahyar Jahangiriesmaili[1], Matthew J. Roorda[1], Chris Bachmann[2], and Rupert Allen[3]

Abstract

Canada signed the Canada–European Union (EU) Comprehensive Economic and Trade Agreement (CETA) in October 2016, which was later ratified by the European Parliament in February 2017. This agreement enables Canadian establishments to trade openly with EU members. Despite the Canadian government's economical analysis, no study has been undertaken to investigate the impact of CETA on Canada's transportation network. The objective of this paper is to assess the potential impact of CETA on the Canadian transportation network by estimating origin–destination trade flows, mode shares, and transportation flows before and after the agreement. Annual provincial commodity flows are obtained from a computable general equilibrium model. Mode shares are determined based on US commodity flow survey data. Finally, changes in freight flows by the rail and truck modes as a result of CETA are examined. The model results indicate an increase in freight movement to and from Eastern Canada and a decrease in trade movement near the US–Canada borders and Western Canada.

Canada has signed several free trade agreements (FTAs) since the North American Free Trade Agreement (NAFTA). The Canada–Israel FTA, the Canada–European Free Trade Association FTA, and the Canada–Korea FTA are only three of 11 FTAs that have been put into place since 1994 (*1*). In October of 2016, Canadian authorities came to an agreement with the EU to sign the Comprehensive Economic and Trade Agreement (CETA). This agreement was approved by the European Parliament in February 2017. The Canadian government believes that CETA reduces tariff barriers and increases accessibility to the EU market (*2*). Potential impacts of CETA on Canada's economy, regulation, society, and employment have been assessed by the Canadian government, but no comprehensive quantitative analysis has been conducted to evaluate Canada's transportation network under the trade agreement (*3*).

Canada and the EU are among the top 10 traders of agricultural products, mining products, manufactured products, chemicals, pharmaceuticals, automotive parts, and clothing (*4*). Execution of CETA may alter the path of exports and imports to and from Canada and the EU, as occurred with NAFTA. Once NAFTA was put into force in 1994, the value of United States (US) exports to Canada inflated by $27.3 billion and the value of imports from Canada to US increased by $21.7 billion in <10 years (*5*). NAFTA also influenced freight demand patterns, transportation mode selections, and route choice preferences. Volume of freight by truck and rail from the US to Canada increased by 51 and 112%, respectively (*5*). Demand at border crossings increased because of the alteration in demand patterns and mode choices. The number of trucks at the Michigan–Windsor border crossing grew by over one million by 1998 (*6*).

The objective of this paper is to forecast commodity movements on the Canadian transportation network once CETA is put into effect. This analysis is intended to help public or private agencies, carriers, and shippers to identify potential needs for infrastructure investment, maintenance requirements, and safety and environmental issues. It may also help authorities to analyze the impacts of policies that

[1]Department of Civil Engineering, University of Toronto, Toronto, ON, Canada
[2]Department of Civil and Environmental Engineering, University of Waterloo, Waterloo, Canada
[3]Department of Canadian Heritage, Government of Canada, Gatineau, Canada

Corresponding Author:
Address correspondence to Mahyar Jahangiriesmaili: mahyar.jahangiriesmaili@mail.utoronto.ca

may be executed in the future to improve transportation before the implementation of the agreement (*7*).

Despite the crucial role of freight transportation in the economy, research to evaluate intercity freight flows on the Canadian transportation system is limited (*8–11*). Unavailability of appropriate datasets and complexity of the problems are two major limitations of analyzing intercity freight flows (*9,10*). In Canada, no freight demand database is available that classifies commodity flows by all freight modes of transport. To overcome this limitation, we use the US commodity flow database (i.e., US freight demand database), since the USA is closely comparable to Canada in terms of economic composition, geographical scale, and network. A limitation of this approach is the exclusion of the influence of underlying factors, such as a country's foreign relations, that are specific to Canada on the freight patterns.

The next section reviews intercity freight models available in the literature, followed by a description of data sources. The modeling framework, results, and conclusions are discussed in the final three sections of the paper.

Literature Review

A variety of modeling approaches has been proposed to represent intercity freight flows. One approach is to develop an integrated multimodal transportation network. Subsequently, origin–destination (OD) commodity movements are assigned to the integrated network under the system optimum or user equilibrium assignment paradigms. Development of an integrated multimodal transportation network is helpful for modeling situations where more than one transportation mode is used to carry commodities from origin to destination (*12*). Since rail and truck are the two dominant modes of freight transportation, these integrated multimodal networks are used for forecasting commodity flows by rail, road, and combination of rail and road (rail–road). The models developed by Guélat, Florian, and Crainic (*9*) and Uddin and Huynh (*11*) are examples of such network assignment models.

Stochastic models estimate the probabilities of a set of outputs (*13*). These models account for uncertainties associated with the freight demand and the decision makers (i.e., shippers). Even though stochastic models are broadly used in mode choice modeling, they require large amounts of data. These data are difficult to obtain in freight transportation because of confidentiality agreements and competition between the firms. Many stochastic models that are found in the literature are econometric models that compute the likelihood of a decision maker selecting an alternative given a set of choices (*14*). Multinomial logit and nested logit models are two common econometric models that are employed in freight transportation (*15*). For example, De Jong and Ben-Akiva (*16*), Maia and Do Couto (*10*), and Rich, Holmblad, and Hansen (*17*) incorporate stochasticity into their models to simulate intercity freight flows.

Friesz, Gottfried, and Morlok (*18*) develop a shipper–carrier model that forecasts commodity movements across a transportation network. Their shipper–carrier model is a bi-level freight model that accounts for the interactions between shippers and carriers. Fernandez, de Cea, and Soto (*8*) further improve the shipper–carrier model and solve for the equilibrium between shippers and carriers over a multimodal network. Ben-Akiva and de Jong (*19*) introduce a novel approach in which commodity generation, commodity distribution, and route assignment are carried out at an aggregate level whereas the mode share process is at a disaggregate level. The model of Ben-Akiva and de Jong (*19*) is known as the "aggregate–disaggregate-aggregate" model.

Similarities and differences exist among the reviewed modeling approaches. The above models, in general, assume that commodity generation is exogenous. A majority of the models concentrate on mode choice and freight flow assignment. Except for the models of Fernandez et al. (*8*) and Friesz et al. (*18*), distributions of commodity flows are assumed to be known. Inclusion of transportation attributes is apparent across all models. Network assignment models account for transportation characteristics, such as travel cost and transportation capacity. However, they fail to consider other attributes including commodity features, accessibility and reliability measures, and shipper and carrier attributes that are considered in stochastic and shipper–carrier models. The main reason for differences in freight flow modeling techniques is the availability of data. For instance, network assignment models are appropriate for situations in which the availability of data is limited. One of the shortcomings of these methods is their deterministic approach. Stochastic models acknowledge the limitation of network assignment models and incorporate variability of model outputs. Shipper–carrier models stochastically allocate OD demands to shippers and deterministically select carriers to convey cargo across the transportation network. Most of the reviewed studies do not apply the models to real-world case studies. Those that have conducted case studies establish assumptions to transform the case studies into problems that are consistent with the developed models. Conversion of freight flows (in mass or monetary units) to shipment (vehicle) flows is theoretically and computationally challenging, and it has only been addressed in the models of Uddin and Huynh (*11*) and De Jong and Ben-Akiva (*16*).

Data

Freight data are limited in Canada and there are no freight transportation surveys that cover all modes of transportation across Canada. The limited availability of freight-related datasets is in part due to a lack of resources dedicated to data collection, competition between shippers or between carriers, and confidentiality agreements (*15,20,21*). This limiting factor lowers the accuracy of freight models and causes the outcomes of freight policies to be sub-optimal (*20*). As collection and organization of datasets are costly, we make use of available data resources to generate a dataset that is compatible with other datasets and applicable to possible future use. Below, we summarize the databases that are employed to model Canadian freight flows.

A dataset, developed by Bachmann (*22,23*), is used to obtain the OD freight flows in Canada. This database outlines the total annual weight and value of Canada's exported and imported commodities with and without CETA. The commodity flow matrices (including with and without CETA) are estimated for the 2015 study year. Trade flows are categorized by commodity classification, province of origin/destination, province of entry/exit, port of clearance, trading partner, and international mode of transport. These flows are generated using a computable general equilibrium (CGE) model, a multi-scale multi-regional input output model, and a value to quantity transformation (*23*). Developing a mode share model that estimates freight flows by mode of transport requires a freight demand database that classifies commodities by mode of transport and commodity type. Since no appropriate Canadian freight demand database is available, the US commodity flow survey (CFS) public use microdata (PUM) is used. The PUM classifies 4,547,661 shipments by area of origin/destination, industry class, domestic mode of transport, shipment value, shipment weight, OD distance, country of export, commodity group, and quarter of 2012 in which the shipment is transported.

Trade flows are assigned onto well-connected transportation networks. Road, rail, marine, and air networks are acquired from the Ministry of Transportation of Ontario. These networks specify the geospatial locations of the infrastructure and outline some of their attributes, such as number of lanes, speed limits, operators, and road or track classifications.

Acquired databases are initially modified to ensure their applicability and compatibility. Allocation of the commodities onto the transportation network requires a greater degree of disaggregation. Therefore, province–province trade flows are disaggregated to economic regions, which are smaller in size. Population (imports) and employment (exports) of the economic regions are used as the basis of disaggregation. Commodities are aggregated based on the Standard Classification of Transported Goods (SCTG) Group that is defined by CFS to satisfy the concordance between the commodity classification of Bachmann (*22,23*) and commodity standard of CFS. A Canada Border Services Agency (CBSA) database is employed to specify the geographical locations, infrastructure types (e.g., border crossing, marine ports, airports, or inland offices), and offered services (e.g., water, rail, road, or air) of the ports of clearance. Since commodity flows are only known for international freight (imports and exports), domestic shipments are removed from the CFS. Extreme outliers, for example unreasonably heavy shipments conveyed using aircraft, are identified and deleted from the survey using the method of boxplot. Availability (transportation accessibility) of the recorded domestic mode of transport is also examined to ensure the accuracy of the dataset. Production and consumption areas and ports of clearance are connected to the road network via a set of direct links as the roadway system is available in all locations. As shipments are transloaded onto the rail network at marine ports or intermodal facilities (i.e., rail terminals), production and consumption zones are connected to the rail network via the nearest intermodal facility and the road network. Ports of clearance that offer rail services are connected to the rail network with a set of arbitrary direct links. Marine ports and airports that are not listed in database of Bachmann (*22,23*) are removed from the network database.

Modeling Framework

We predict the domestic mode of transport for each commodity flow based on the services that are provided by the ports of clearance for that flow. For instance, if the port of clearance is a highway border crossing or an airport, we assume that the domestic mode of transport is truck. We specify the services that are provided by each port of clearance based on their offered services (defined by CBSA), geographical location, and international mode of transport. Notably the model does not include the pipeline mode as the data for shipments by pipeline is not available in the US CFS database.

We assume that a port of clearance can provide one or two types of services, and the offered services are air, rail, road, or marine. Ports of clearance that process air shipments are most likely to be airports. Since the majority of airports have road access and aircraft carry light-weight and low-volume goods, truck access is assumed for shipments that are cleared by these ports. Ports of clearance that offer railway services are rail depots located on US–Canada borders. Railways are accessed via roadways at domestic intermodal facilities. Therefore, the assumed domestic of mode of transport for rail depots is truck–rail intermodal. Ports of clearance that provide roadway services are US–Canada roadway border crossings. Their assumed domestic mode of transport is truck. Domestic mode of transportation may be truck or rail for ports of clearance that clear marine cargo. The reason is that marine ports are accessible by truck and rail, and commodities that are cleared by these ports are not restricted by their weight or volume. This assumption ignores occasions where a shipment is loaded onto a vessel at a Canadian harbor and transloaded onto a different vessel at another domestic port. We develop a cross-classification model for this case based on the modified US CFS microdata for US exports to estimate the mode share between rail and truck. This model forecasts the likelihood of selecting the truck mode or the rail mode based on the SCTG Group and the routed distance between the economic regions and the ports of clearance. Table 1 illustrates the cross-classification model in which the likelihood of each class is determined by the annual ton. Considering the assumption made about railway accessibility, rail represents truck–rail intermodal in the cross-classification model.

Table 1. Cross-Classification Mode Share Model for Ports of Clearance that Offer Marine Services

Likelihood: Annual tonne (%)	Routed distance (km)											
	0–250		250–500		500–750		750–1000		1000–1750		>1750	
SCTG group	Rail	Truck	Rail	Truck	Rail	Truck	Rail	Truck	Rail	Truck	Rail	Truck
A	0.1	99.9	14.7	85.3	18.3	81.7	53.3	46.7	17.9	82.1	31.7	68.3
B	1.5	98.5	15.0	85.0	16.2	83.8	24.8	75.2	40.3	59.7	61.9	38.1
C	36.2	63.8	7.1	92.9	62.9	37.1	58.0	42.0	72.1	27.9	74.0	26.0
D	1.4	98.6	20.0	80.0	15.2	84.8	31.1	68.9	22.8	77.2	62.0	38.0
E	8.8	91.2	19.2	80.8	58.5	41.5	51.3	48.7	31.3	68.7	55.4	44.6
F	3.7	96.3	18.4	81.6	12.8	87.2	30.5	69.5	33.8	66.2	27.8	72.2
G	0.4	99.6	15.1	84.9	9.3	90.7	13.1	86.9	28.4	71.6	19.5	80.5
H	8.9	91.1	17.5	82.5	6.5	93.5	6.2	93.8	21.8	78.2	20.7	79.3
I	6.3	93.7	35.7	64.3	6.1	93.9	8.0	92.0	30.5	69.5	40.8	59.2

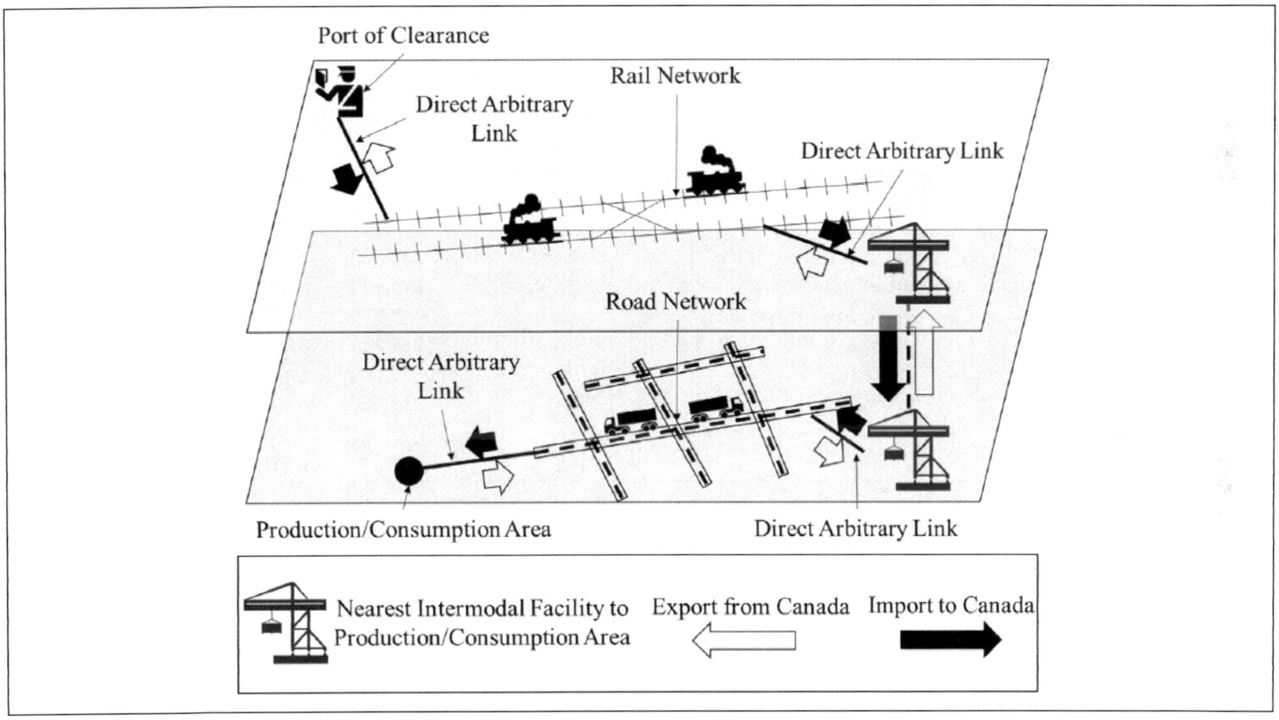

Figure 1. Route assignment of rail cargos over the integrated rail–road network.

Domestic mode of transport for ports of clearance that provide types of services is similarly determined knowing the offered services and the international mode of transport, which is defined by Bachmann (*22,23*).

This methodology narrows the choice set of transportation modes down to rail and road. Modeling intercity freight flow using this approach is reasonable because 93% of total Canadian freight flows are carried by truck or train (*24*). For commodities that are transported by truck, we use the well-connected road network to assign the exports from the points of production to the ports of clearance and model the imports from the ports of clearance to the points of consumption. Assignment of rail cargo onto the rail network requires the identification of intermodal facilities where commodities are transferred between truck and train. Since it is economical to convey rail shipments on dedicated rail tracks, shipments are assumed to be transferred at the intermodal facility that is closest to the points of production or consumption. We use the method of Southworth and Peterson (*25*) to route rail shipments over an integrated truck–rail network. Route assignment of rail exports and imports is demonstrated in Figure 1. Due to the scale of the acquired OD flows, annual freight volumes are assigned onto the rail and road network in TransCAD 4.5 using all-or-nothing assignment.

Table 2. Changes in Commodity Flows by Domestic Mode and Province of Production/Consumption for Exports and Imports as a Result of CETA

Annual weight (tonnes)			Difference (Tonnes) Rail	Difference (Tonnes) Truck	Difference (%) Rail (%)	Difference (%) Truck (%)
Exports	Province of Production	Alberta	−66,700	−485,500	−0.7	−0.5
		British Columbia	−291,100	−639,000	−0.8	−0.7
		Manitoba	−20,600	−58,500	−1.0	−1.0
		New Brunswick	−24,700	−65,400	−1.1	−0.5
		Newfoundland and Labrador	−207,800	−146,700	−0.7	−0.6
		Nova Scotia	−87,100	−362,700	−0.7	−0.6
		Ontario	−12,400	−545,100	−0.2	−0.7
		Prince Edward Island	0	1,500	0.6	0.6
		Quebec	−129,000	−163,300	−0.8	−0.5
		Saskatchewan	22,000	−3,600	0.2	0.0
		Yukon, Northwest Territories, Nunavut	−200	−400	−0.7	−0.7
	Total		−817,600	−2,468,700	−0.7	−0.6
Imports	Province of consumption	Alberta	78,100	136,200	2.1	1.0
		British Columbia	100,600	116,300	2.3	0.9
		Manitoba	21,500	31,600	2.1	1.0
		New Brunswick	5,700	22,300	0.5	0.6
		Newfoundland and Labrador	6,900	15,800	1.5	0.9
		Nova Scotia	7,600	22,200	1.1	0.9
		Ontario	212,200	627,200	1.2	1.1
		Prince Edward Island	800	2,500	2.0	1.5
		Quebec	26,700	269,200	0.4	1.0
		Saskatchewan	21,200	31,100	2.5	1.0
		Yukon, Northwest Territories, Nunavut	4,700	5,900	1.9	1.3
	Total		485,900	1,280,200	1.3	1.0

Results

The above modeling framework is applied with the database of Bachmann (*22,23*) and the outcomes of the model are discussed in this section. Table 2 summarizes the change in commodity flows by province by domestic mode of transportation for exports and imports once CETA is enforced. Tonnage of exports overall decreases by approximately 0.6%. The reason for the decrease is that larger tariffs on higher value goods are removed with CETA. As a result, industries producing higher value goods expand and bid up prices of factors (land, labor, and capital). Industries producing lower value goods become less attractive and cannot compete for the factors of production that are now more expensive. Therefore, industrial composition is shifted toward higher value goods, therefore, total export volume decreases, even though the total value of exports increases.

The decrease in export tonnage occurs across all provinces with the exceptions of Prince Edward Island and Saskatchewan. The largest absolute export reductions for rail originate in Newfoundland and Labrador and British Columbia. The largest absolute export reductions by the truck mode originate in British Columbia, Ontario, and Alberta. Imports increase for all provinces with a nationwide average increase of 1.3% for the rail mode and 1.0% for the truck mode. With CETA, Ontario imports increase the most, in absolute tonnage, for both truck and rail. The greatest percentage increases in rail commodity imports are destined for Saskatchewan, British Columbia, Manitoba, Prince Edward Island, and Alberta. The greatest percentage increases for the truck mode are destined for Prince Edward Island and the Northern Territories, although the absolute increases are small.

Table 3 outlines the impacts of CETA on commodity flows by commodity group and by domestic mode of transportation for exports and imports. The reduction in export tonnage by both rail and truck is dominated by reductions in the export of commodity group C (stone, non-metallic minerals, and metallic ores). Significant reductions also occur for the truck mode in commodity group D (Coal and petroleum products) and F (Logs, wood, textiles, and leather). However, the reduction for commodity group D should be

Table 3. Changes in Commodity Flows by Domestic Mode and SCTG Commodity Group for Exports and Imports as a Result of CETA

Annual weight (tonnes)			Difference (tonnes) Rail	Difference (tonnes) Truck	Difference (%) Rail (%)	Difference (%) Truck (%)
Exports	SCTG group	A	78,900	52,000	1.0	0.2
		B	2,000	79,000	0.1	1.0
		C	−579,000	−1,235,000	−0.7	−0.7
		D	−19,100	−511,500	−0.3	−0.4
		E	−58,600	−175,700	−0.8	−0.8
		F	−230,300	−510,200	−1.7	−1.3
		G	−15,300	−168,900	−1.2	−1.1
		H	3,300	2,600	3.7	0.0
		I	400	−900	4.2	−0.4
	Total		−817,600	−2,468,700	−0.7	−0.6
Imports	SCTG group	A	13,100	50,000	0.9	0.7
		B	134,500	24,000	5.6	0.3
		C	157,600	185,500	2.0	2.0
		D	5,400	2,300	0.1	0.0
		E	−49,100	−151,200	−0.6	−0.8
		F	66,800	307,200	1.2	2.0
		G	128,600	728,500	2.9	2.1
		H	26,700	127,800	5.0	0.9
		I	2,400	6,000	3.0	1.3
	Total		485,900	1,280,200	1.3	1.0

treated with caution because the pipeline mode is not accounted for in this mode share model. Imports increase with CETA across all commodity groups with the exception of Group E (Basic chemicals, chemical, and pharmaceutical products). The largest tonnage increases occur in Group G (Base Metals and Machinery), and Group F (Logs, wood products, textiles, and leather) by the truck mode. The greatest absolute increases in import tonnage by rail are for commodity group C (Stone, non-metallic minerals, and metallic ores), Group B (Grains, alcohol, and tobacco products), and Group G (Base metals and machinery).

Demand for ports of clearance that are located on the east coast of Canada is estimated to increase following the trade agreement with the EU. For example, annual tonnage of exports from the port of Montreal is expected to increase by 2.3% by rail and 3.2% by truck. Volume of exports through the ports of Vancouver, Hawkesbury, and Sept-Îles are estimated to decline by 1.1, 0.4, and 0.3 million annual tonnes, respectively. These ports are expected to encounter the highest reduction in exports by rail and truck. On the other hand, volume of exports via the ports of Montreal and Baie-Comeau are anticipated to grow by 0.7 million annual tonnes. The volume of imported goods is projected to grow by 1.6 and 0.4 million annual tonnes at the ports of Montreal and Halifax. This growth increases the truck and rail movements by 10 and 9% at the Port of Montreal and the Port of Halifax. The largest decreases in import tonnage are by the truck mode at the Ambassador Bridge (Windsor), Sarnia, and Fort Erie (Niagara). The volume of truck freight from these ports is forecasted to be reduced by 0.4 million annual tonnes.

Bachmann (*3*) notes that CETA diverts the commodity movements away from Canada–US border and shifts it toward the Atlantic region. The reduction in trade with the US is known as "trade diversion." This term was first introduced by Viner (*26*), who declares that trades tend divert from a trade partner on enforcement of a new trade agreement with another partner. This complies with the negative changes in commodity flows (especially for exports) that can be seen in Tables 2 and 3, as rail and truck are the preferred modes for conveying trans-border freight. To fully observe the changes in trade patterns, we map the results from the route assignment using Geographic Information System software. This process is undertaken by first mapping out the trade flows of the scenarios before and the after the CETA agreement and then comparing the generated maps against one another. Outcomes of the comparison are shown in Figures 2 and 3. Figure 2 illustrates the changes in truck freight flows (in millions of annual tonnes) and Figure 3 displays the changes in rail freight flows (in millions of annual tonnes). One of the outcomes of this comparison indicates that the total increase in freight volume as a result of CETA is 609 million tonne-km for rail (+0.3%) and 1139 million tonne-km (+0.2%) for truck. Mapped results of the before agreement demonstrate that freight flows are more significant along the Quebec City–Windsor Corridor, US–Canada border, and roadways or railways that connect the

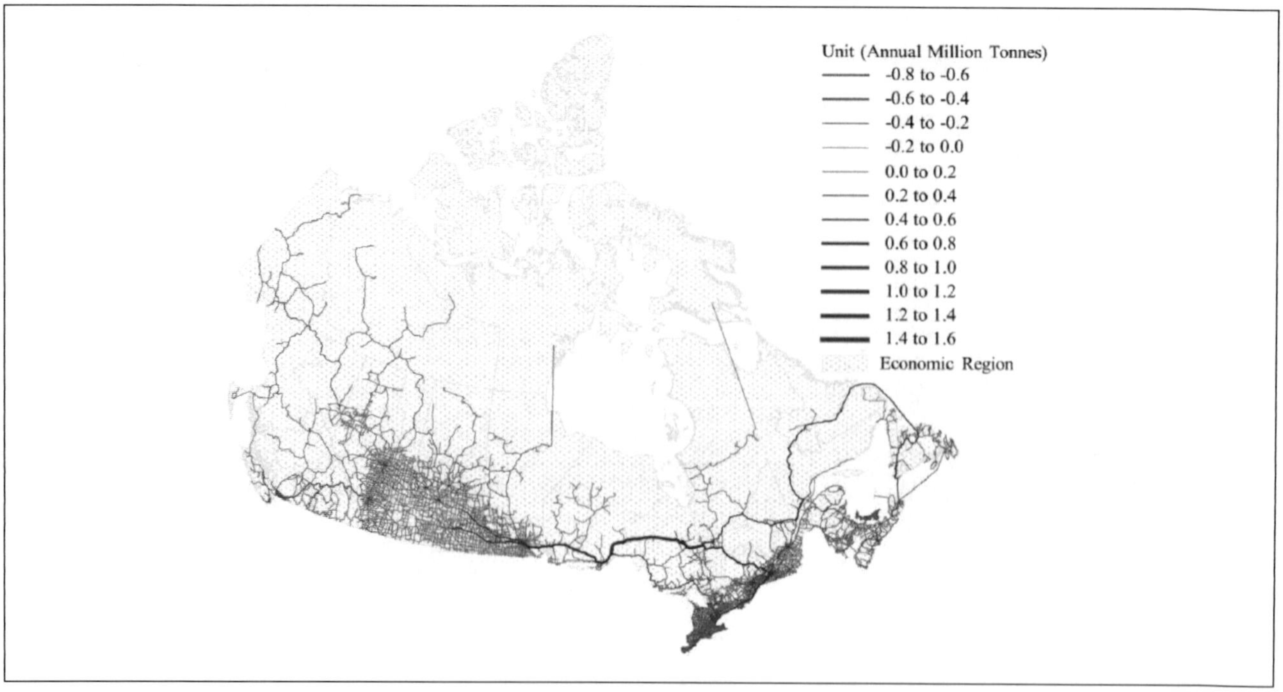

Figure 2. Changes in annual imports and exports on the Canadian highway network as a result of CETA.

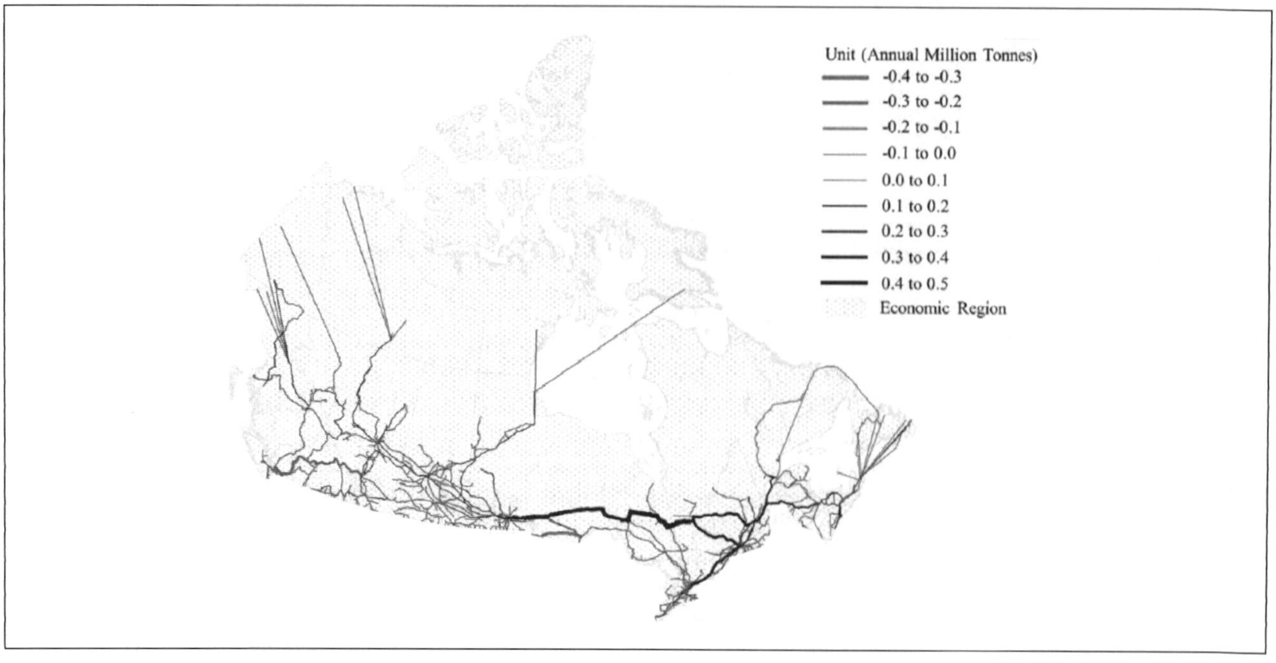

Figure 3. Changes in annual imports and exports on the Canadian rail network as a result of CETA.

east coast to the west coast. Moreover, our analyses specify that volumes of exports and imports along roadways and railways along the Quebec City–Windsor Corridor and Central Canada are expected to grow even more once CETA is put into place. Implementation of the agreement also reduces the commodity movements via west coast ports and US–Canada border crossings as Canadian shippers and carriers prefer Atlantic gateways for transporting goods.

Conclusion

This paper provides a preliminary estimate of the change in domestic freight flows by mode of transport associated with changes in import and exports after the implementation of CETA. Three freight datasets have been used: a commodity flow matrix generated from a CGE model, the US CFS, and transportation network data. After data adjustments to ensure consistency, international freight flows are divided between the road and rail modes based on the characteristics of the ports of clearance and the US CFS mode share distributions. Trades flows are then assigned onto road and rail networks to analyze the impacts of the CETA on freight patterns.

The model results indicate that the demands for both rail and road (in tonnage) are expected to increase once the CETA is put into effect. This increase occurs across the Eastern provinces as shippers and carriers export and import their shipments to/from Europe via the Atlantic gateways. Although the percentages increases may appear to be small, these translate to very real strains on the system in the Atlantic region, where an increase in train and truck loads have the potential for a very real impact for modes near capacity and with urban constraints. As infrastructure and trade networks in Canada have been predominately focused toward the US border and western ports, it is unclear how quickly the economy can pivot in this direction, and what that would mean for the larger transportation network. Whereas benefits to the Atlantic region may result from the CETA agreement, for these to be fully realized, it is likely that transportation infrastructure investment would need to be made.

The shift in freight reduces commodity movements near the US–Canada border and the Western provinces. Moreover, even though traffic volumes through the roadways and railways across the Quebec City–Windsor Corridor are currently high, the demand for this part of the transportation network is expected to grow. A reduction in exports associated with CETA is expected for both truck and rail modes for all provinces except Prince Edward Island and Manitoba. Imports increase for shipments to all provinces by both the rail and truck mode and for all commodity groups except for basic chemicals, chemical and pharmaceutical products. The greatest absolute increase by weight is for base metals and machinery by the truck mode. The greatest decrease in exports is predicted to occur at the Port of Vancouver port of clearance, and the greatest increase in imports is predicted for the Montreal port of clearance.

Despite the importance of freight transportation to the economy, limited access to freight data constrains the modeling process. Some of the main limitations experienced in this research because of the data restrictions are listed below.

I. The mode shares for domestic freight flows are based on US CFS data, which is considered to be a reasonable surrogate for Canadian mode shares.

II. The pipeline mode of transportation has been ignored. This likely results in an overestimate of domestic commodity flows by truck and rail of commodity group D.

III. The concordance between GSC-2 and SCTG-2 commodity classification systems is imperfect.

IV. Little information about the services offered at inland office ports of clearance was available from public CBSA websites. Assumptions were required regarding which import/export modes of transportation are serviced from these offices.

Improved access to freight data is essential to improve understanding of the underlying structure of the freight transportation system. This is to allow trade agreements to be fully optimized, and to ensure that countries are well placed to take advantage of increased trade. The problem is not that the data are not collected, but that they are not publically disseminated. Freight data are largely held privately by freight operators, preventing use for modeling, analysis, system optimization, and shared approaches to joint problems. The siloed nature of Canadian transportation data means that only simplified approaches can be used to inform economic projections, and to inform key trade policy decisions. This drawback can be resolved to some extent by allocating more resources to freight surveys and strengthening the collaboration between the public and private agencies. Improved information is crucial when governments are making large-scale transportation infrastructure investments.

As nations look to global markets to grow domestic companies and secure foreign investment, it is increasingly essential to understand not just the increase or decrease in exports or imports resulting from a new FTA, but what the impacts to the freight transportation system will be and whether the system can handle the increased load across all modes. Many freight transportation systems are nearing capacity, notably ports, and port rail or truck intermodal connections and increased load in either direction could result in costly delays, and blockages. Determining the capacity and potential for handling increased freight demand as a result of CETA is an area for future research, providing value to trade policy makers and infrastructure investment decision makers.

Acknowledgments

This project is funded by Innovation, Science and Economic Development Canada, formerly Industry Canada.

References

1. Government of Canada. Trade and Investment Agreement. 2017. https://www.international.gc.ca/trade-commerce/trade-agreements-accords-commerciaux/agr-acc/index.aspx?lang=eng. Accessed May 1, 2017.
2. Government of Canada. Agreement Overview. 2016. http://www.international.gc.ca/trade-commerce/trade-agreements-accords-commerciaux/agr-acc/ceta-aecg/overview-apercu.aspx?lang=eng. Accessed Mar. 1, 2017.

3. Bachmann, C. Modeling the Impacts of Free Trade Agreements on Domestic Transportation Gateways, Corridors, and Ports. *Transportation Research Record: Journal of the Transportation Research Board*, 2017. 2611: 1–10.
4. WTO. World Trade Statistical Review. 2016. https://www.wto.org/english/res_e/statis_e/wts2016_e/wts16_chap9_e.htm. Accessed May 1, 2017.
5. Chi, J. Assessing Long-Run Determinants of Cross-Border Freight Flows Between the United States and Canada. Presented at 93rd Annual Meeting of the Transportation Research Board, Washington, D.C., 2014.
6. Club, S., S. Holbrook-White, and T. C. Fund. *NAFTA Transportation Corridors: Approaches to Assessing Environmental Impacts and Alternatives*. Presented at North American Symposium on Understanding the Linkages between Trade and Environment, Washington, D.C., 2000.
7. Woudsma, C. NAFTA and Canada–US Cross-Border Freight Transportation. *Journal of Transport Geography*, Vol. 7, No. 2, 1999, pp. 105–119.
8. Fernandez, J. E., J. de Cea Ch, and A. Soto. A Multi-Modal Supply–Demand Equilibrium Model for Predicting Intercity Freight Flows. *Transportation Research Part B: Methodological*, Vol. 37, No. 7, 2003, pp. 615–640.
9. Guélat, J., M. Florian, and T. G. Crainic. A Multimode Multiproduct Network Assignment Model for Strategic Planning of Freight Flows. *Transportation Science*, Vol. 24, No. 1, 1990, pp. 25–39.
10. Maia, L. C., and A. F. Do Couto. An Innovative Freight Traffic Assignment Model for Multimodal Networks. *Computers in Industry*, Vol. 64, No. 2, 2013, pp. 121–127.
11. Uddin, M. M., and N. Huynh. Freight Traffic Assignment Methodology for Large-Scale Road–Rail Intermodal Networks. *Transportation Research Record: Journal of the Transportation Research Board*, 2015. 2477: 50–57.
12. Bontekoning, Y. M., C. Macharis, and J. J. Trip. Is a New Applied Transportation Research Field Emerging?: A Review of Intermodal Rail–Truck Freight Transport Literature. *Transportation Research Part A: Policy and Practice*, Vol. 38, No. 1, 2004, pp. 1–34.
13. Taylor, H. M., and S. Karlin. *An Introduction to Stochastic Modeling*, Academic Press, California, 1994, pp. 1–17.
14. McFadden, D. Conditional Logit Analysis of Qualitative Choice Behaviour. In *Frontiers in Econometrics* (P. Zarembka, ed.), Academic Press, New York, 1973, pp. 105–140.
15. De Jong, G., H. Gunn, and W. Walker. National and International Freight Transport Models: An Overview and Ideas for Future Development. *Transport Reviews*, Vol. 24, No. 1, 2004, pp. 103–124.
16. De Jong, G., and M. Ben-Akiva. A Micro-Simulation Model of Shipment Size and Transport Chain Choice. *Transportation Research Part B: Methodological*, Vol. 41, No. 9, 2007, pp. 950–965.
17. Rich, J., P. M. Holmblad, and C. O. Hansen. A Weighted Logit Freight Mode-Choice Model. *Transportation Research Part E: Logistics and Transportation Review*, Vol. 45, No. 6, 2009, pp. 1006–1019.
18. Friesz, T. L., J. A. Gottfried, and E. K. Morlok. A Sequential Shipper-Carrier Network Model for Predicting Freight Flows. *Transportation Science*, Vol. 20, No. 2, 1986, pp. 80–91.
19. Ben-Akiva, M., and G. de Jong. The Aggregate-Disaggregate-Aggregate (ADA) Freight Model System. In *Freight Transport Modelling*, Emerald Group Publishing Limited, 2013, pp. 69–90.
20. Tavasszy, L., and G. De Jong. Comprehensive Versus Simplified Models. *Modelling Freight Transport*, Elesvier, London, 2014, pp. 245–256.
21. Albert, A., and A. Schafer. Demand for Freight Transportation in the US: A High-Level View. *Journal of Transportation Statistics*, 2013, pp. 103–116.
22. Bachmann, C. *Analyzing the Transportation Infrastructure Impacts of Trade Agreements*, Presented at Canadian Transportation Research Forum 51 Annual Conference, Toronto, ON, 2016.
23. Bachmann, C. *Analyzing the Infrastructure Impacts of Free Trade Agreements*. Presented at 19th Annual Conference of Global Economic Analysis, Washington, D.C., 2016.
24. Transport Canada. Transportation in Canada 2015. 2015. https://www.tc.gc.ca/media/documents/policy/2015_TC_Annual_Report_Overview-ENAccessible.pdf. Accessed Dec. 20, 2016.
25. Southworth, F., and B. E. Peterson. Intermodal and International Freight Network Modeling. *Transportation Research Part C: Emerging Technologies*, Vol. 8, No. 1–6, 2000, pp. 147–166.
26. Viner, J. *The Customs Union Issue*, Carnegie Endowment for International Peace, New York, 1950, pp. 1–34.

The Standing Committee on International Trade and Transportation (AT020) peer-reviewed this paper (18-01611).

Article

TRR
JOURNAL OF THE TRANSPORTATION RESEARCH BOARD

Transportation Research Record
2018, Vol. 2672(11) 19–29

DOI: 10.1177/0361198118774234
journals.sagepub.com/home/trr
SAGE

Geospatial Risk Assessment of Marine Terminal Infrastructure to Storm Surge Inundation and Sea Level Rise

George M. McLeod[1], Thomas R. Allen[2], and Joshua G. Behr[3]

Abstract

Planning resiliency and sustainability of port operations and critical infrastructure requires risk assessment of storm surge exposure and potential sea level rise. An approach for rapid, screening-level assessment is developed to estimate the current and future risk of exposure to severe storm surges posed to marine terminal facilities in Norfolk, Virginia. The approach estimates the vertical elevation of local mean sea level fifty years into the future and attendant increases in potential storm surge heights. Inundation models are designed for baseline water levels and storm surges for category 1–3 hurricanes across five precautionary future sea level rise scenarios. In addition, tidal flooding poses an emerging threat because sea level rise will also force tides to higher elevations, suggesting that today's extreme high tides may be the future mean high tide and today's "nuisance" tidal flooding may in the future recur with chronic regularity. Potential tidal flooding levels are also modeled for each sea level scenario. This approach allows a port to assess relative risk tolerance across the range from lesser to more severe flooding events. Maps and tabular information in linked scenarios are used to summarize the extent, pattern, and depth of potential flooding. The methodology and data developed in this study may be applied to inform the timing and placement of planned assets and can be leveraged in the broader pursuit of optimization in support of long-term master planning at marine terminals.

Old Dominion University has developed a robust program of research and engagement with local, regional, national, and international groups working to address increasing coastal hazards posed by sea level rise (SLR). Prompted by recent planning activity at the Port of Virginia, the research team sought the most recent scientific information from several studies by federal agencies focused on the Hampton Roads port region of southeastern Virginia. This background phase compiled observations and trends from these studies and synthesized peer-reviewed literature relevant to vulnerability assessment, modeling, and risk management in ports across the world.

Numerous SLR vulnerability studies have been conducted in or near the region. However, none has specifically evaluated site and infrastructure exposure to storm surge hazard and SLR in the marine terminals of the Port of Virginia. The US Army Corps of Engineers' study (Burkes-Copes et al.) (*1*) of the Norfolk Naval Station required several years and millions of dollars to evaluate systemic risks and provides the nearest analog with the nearby Port of Virginia marine terminal, both spatially and regarding specific facilities. Further, Titus and Cacela (*2*) implemented an analysis of regional vulnerability, although this was conducted at a scale too coarse for site-level inference. To the south, the North Carolina Sea Level Rise Risk Management Study (*3*) is a robust regional modeling assessment, but the focus on flood zones and future risks posed to the State and National Flood Insurance Program is markedly different than the concerns of a marine terminal. Other relevant studies have either tended to focus on economic and property vulnerability (*4*), specific sectors such as metropolitan public roadways, or utilized bathtub models lacking hydrologic connectivity or ambiguous tidal flooding (*5*), use static incremental water level changes (*6*), or are limited in spatial extent and temporally focused on near real-time prediction rather than long-term SLR (*7*).

As a case study, the objective was to answer two fundamental questions on the exposure of the Port of Virginia's Norfolk International Terminals South (NITS) facility to future storm surge flooding concomitant with relative SLR:

[1]Department of Ocean, Earth, and Atmospheric Sciences, Old Dominion University, Norfolk, VA
[2]Department of Political Science & Geography and Commonwealth Center for Recurrent Flooding Resiliency, Old Dominion University, Norfolk, VA
[3]Virginia Modeling, Analysis, and Simulation Center, Old Dominion University, Suffolk VA

Corresponding Author:
Address correspondence to George M. McLeod: gmcleod@odu.edu

(1) What is the magnitude and spatial extent of potential surge-related flooding that may be expected at the terminal, both currently and into the future, given SLR scenarios?; and (2) What is the current and future vulnerability, in terms of flood risk exposure, of the structural assets to this surge-related flooding under these given SLR scenarios? In lieu of presenting the individual analyses and results of the ports facilities, this paper focuses on the methods developed to offer guidance beyond this specific case, facilitate study of other regional port facilities, and provide pathways for necessary adaptation and mitigation actions pertinent to planning, sustainable design, and facility resiliency.

Literature Review

Prior SLR Studies

The Hampton Roads' Sea Level Rise (HRSLR) Preparedness and Resilience Intergovernmental Pilot Project (*8*) recommended that the region foster improved ways for agencies to work together and the adoption of regional sea level planning scenarios and standards. The region has not adopted regional sea level planning scenarios and standards upon which this case study may rely for guidance. The NITS project, however, does leverage recent related efforts by SERDP (*9*) "Regional Sea Level Scenarios for Coastal Risk Management: Managing the Uncertainty of Future Sea Level Change and Extreme Water Levels for Department of Defense Coastal Sites Worldwide," and two recent studies by the National Oceanic and Atmospheric Administration (NOAA) focused on regionalizing sea level trends (*10*) and preparing for future non-linear increases in "nuisance," "sunny-day," or "recurrent" tidal flooding (*11*). Key highlights from these studies include the HRSLR Pilot study's recommendation that critical infrastructure vulnerability to SLR impacts focus on the next 30, 50, to 75 years. The same group noted interdependencies between private and public infrastructure systems, an externality that bears consideration for this project, which does not directly evaluate impacts on City of Norfolk or Navy properties and their infrastructure and transportation systems with significant connection to NITS. From outside the region, some specific best management practices can be drawn from Southeast Florida and New Orleans relevant to the Port, including:

- Strive for regionally consistent mapping methods and products for SLR planning.
- Develop consistent criteria for risk assessments.
- Evaluate management strategies for storm water and flood control/drainage structures.
- Include tidal flooding in risk assessment studies.

Port Response to Climate Change Risks

In reviewing the state of scientific study on ports relative to SLR, trends are evident in sectors and scale of analysis. There exists a core, traditional body of research which focuses on ports and their regional to global-scale function. There is also a growing body of literature on the site-specific dimensions of port vulnerability to SLR. Relatively few studies scale between these to the level of whole facilities or regions, creating somewhat of a knowledge gap on what policies and best practices port planners and governing authorities can exchange and implement.

Some studies of individual ports have tended to focus on the ports in the context of urban metropolitan systems with vulnerabilities including infrastructure, transportation, economics, and social vulnerability. Akukwe and Ogbobo (*12*), for example, analyzed Port Harcourt, Nigeria, evaluating exposure of the port's economic and social vulnerabilities to create indices for tracking. Also working at the city scale, Hallegatte et al. (*13*) focused on identifying economic impacts of SLR and storm surges in Copenhagen, cautioning that predictions in changes in storminess were indeterminate for Copenhagen.

At the coarser, global scale, Becker et al. (*14*) utilized social science surveys of port administrators to investigate preparedness for climate change. The study revealed that many ports acknowledge SLR and climate-related issues, but have yet to address them.

Hanson et al. (*15*) took a similar, global-scale approach to evaluating port cities and climate extremes. They estimated that about 40 million people are now exposed to the threat of 1:100 year coastal flood events, and that population could triple by the 2070s owing to combined population growth in coastal cities and relative sea level change. In fact, Norfolk-Virginia Beach was noted as one of the global top-ranked metro areas (#19 globally in assets exposed by the 2070s, rising from $84.8B today to $581.9B in the future).

The Port of Miami, The Port Authority of NY and NJ, the Massachusetts Port Authority, Port of Seattle, Port of Corpus Christi, and Georgia Ports Authority are each mentioned for various initiatives. Nonetheless, common challenges were revealed, such as the region-specific climate information needs of ports requiring more data and research. Nicholls (*16*) ranked the exposure and vulnerability of global ports to climate extremes. In summary, these coarser-scale studies suggest overarching issues for consideration by port administrators at the local level:

- Short- and long-term plans should benchmark climate change forecasts as the science improves, such as an interval of 5 years (with projections out 50 years).
- Port administrators should keep abreast of scientific developments, trends, and technologies for potential adaptation.
- Ports ought to ally with regional entities facing the similar challenges.
- Coordinated studies, such as state, regional, and multi-nodal nature, would be fruitful.

At the site level, port literature is often focused on engineering design in the face of increasing wave energy and flood impacts. Numerical modeling is demonstrated by Rajabalinejad and Demibilek (*17*) for engineering structures exposed to SLR. They recommend modeling for the capacity to inform uncertainties with current practices and designs for coastal flood protection. Finally, some ports exposed to wave overtopping have demonstrated that scenario-based approaches are useful analyses (*18*). From a coastal risk management perspective, Sierra et al. conclude that it may be best to utilize lower probability, extreme (upper tail end) events rather than Intergovernmental Panel on Climate Change scenarios (*19*).

With specific relevance to this NITS study, Good et al. (*20*) noted that local participants did not use GIS-based maps to the extent hoped, depending rather upon local experts, technical advisors, and aerial photo and static maps (although they do cite the value of GIS.) Overall, Good et al. recommended early engagement with a diversity of port stakeholders and capturing a wider range of community vulnerability.

Sea, Lake, and Overland Surges from Hurricanes (SLOSH) Model

Storm surge modeling data may be obtained from the SLOSH Display Program and used to map the extent and depth of inundation resulting from various storm surge levels with SLR. The SLOSH model was developed by the National Weather Service National Hurricane Center (NHC) to "…estimate storm surge heights, resulting from historical, hypothetical, or predicted hurricanes by taking into account the atmospheric pressure, size, forward speed, and track data" (*21*).

Composite storm surge modeling methods are available within SLOSH, specifically a Maximum of MEOWs (MOMs) approach "which are regarded by the NHC as the best approach for determining storm surge vulnerability for an area since it takes into account forecast uncertainty" (*21*). This method incorporates storm simulations from thousands of model runs with the same category, forward speed, storm trajectory, and initial tide level. MOMs provide a worst-case scenario product as they are compiled based on the maximum storm surge height for all hurricanes of a given category. When assessing storm surge impacts, utilization of SLOSH MOMs for each category (1–5) storm occurring at high tide is most effective for identification of infrastructure vulnerability as a worst-case scenario for risk identification (*22*).

Relative SLR Scenarios

Several studies note the need to consider future scenarios for SLR, considering a wide range of potential changes, and also an emphasis on prudent, conservative estimates owing to plausible consequences of ice sheet degradation, localized uncertainty in subsidence, and retaining "no-regrets" with regard to possible long-term adaptation. Sweet et al. (*10*) recommend using a scientifically plausible upper bound (worst-case or extreme scenario) in addition to intermediate scenarios. Although lower probability, using higher forecasts reveals overall system risks and information for long-term strategies. A complementary mid-range scenario is appropriate for shorter-term planning. The two, in combination, can be considered as a "general planning envelope" (*10*). Accordingly, we reproduce the sea level curves for potential consideration of the adopted scenarios (Figure 1).

Approach and Methods

SLR Scenario Selection

To provide context for inundation maps and projected SLR and surge in time, we utilized the USACE Sea Level Rise Curve Calculator (*23*). The study includes inundation maps for the following four relative SLR scenarios for storm surge model simulations and tidal "nuisance" flooding with approximate years of potential realization, with the first year reflecting the fastest acceleration of SLR (NOAA extreme) and the latest the NOAA intermediate-high: 7.9" (20 cm), 2022–2032; 15.7" (40 cm), 2035–2045; 23.6" (60 cm), 2045–2058; 31.5" (80 cm), 2055–2072.

Study Area

The boundary of the study area includes the Port of Virginia's NITS facility and immediately adjoining areas (Figure 2). The South Terminal comprises 272 acres of NIT, which is the Virginia Port Authority's largest terminal.

Assumptions

In keeping with conservative approach for risk identification, we assume no engineering intervention or related landform changes that would mitigate storm surges under relative SLR. Similarly, we assume no major shoreline modifications or fortifications around the facility.

Elevation Data. This study's variety of data incurs potential error and uncertainties, which we have endeavored to detail and mitigate data error and uncertainty as best possible. These include potential error in the digital elevation model (DEM), which we evaluate using hydro-connected storm surge simulation and Monte Carlo (MC) simulation.

Gesch (*24*) stipulated that input elevation information is a primary contributor to the uncertainty associated with inundation hazard assessments and that, because these data are such a critical component in coastal hazard assessments, the vertical accuracy strongly influences the reliability of the results. For SLR assessments, Gesch (*25*) suggested that utilization of "best available" light detection and ranging (LiDAR) data having both high spatial resolution and accuracy is preferable. A key descriptive metric of the accuracy of

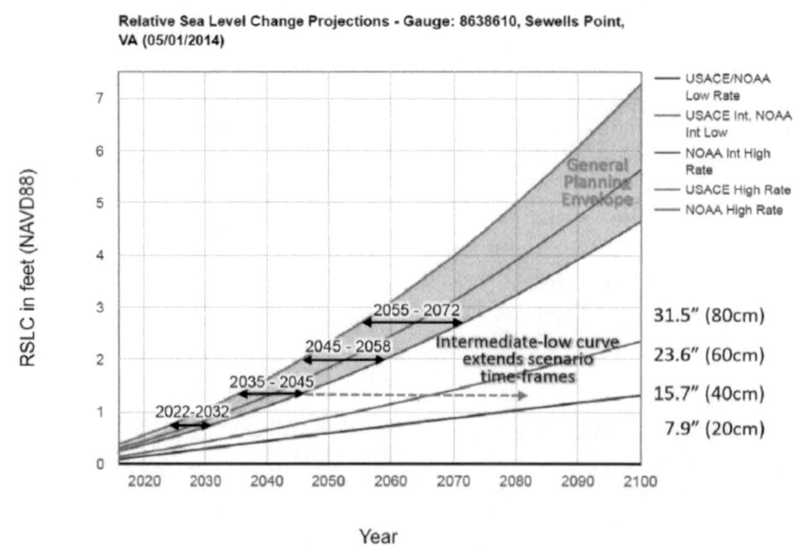

Figure 1. Relative sea level projection curves for Sewell's Point, VA using US Army Corps of Engineers Sea-Level Change Calculator. NOAA and USACE curves, general planning envelope between intermediate-high and extreme curves (orange fill) are shown. Arrows depict time ranges for relative sea level scenarios within the planning envelope between NOAA High and Intermediate-High SLR that were used in modeling storm surges and tidal flooding.

Figure 2. Norfolk International Terminal South (NITS) study area location.

digital elevation data is fundamental vertical accuracy, which describes vertical accuracy at the 95-percent confidence level in open terrain where errors should approximate a normal error distribution (*26*). The fundamental vertical accuracy for the current best available LiDAR data for the NITS terminal study region used for this study was reported as ± 0.129 m (*27*).

Model Error and Uncertainty. The NOAA SLOSH model contains inherent uncertainty owing to the coarse resolution of the model grid. In a conservative approach, this research rounded interpolated values to the highest adjoining SLOSH cells for areas that were not initially characterized as flooded.

The impact of positional error in the LiDAR DEM was considered during the analysis of the potential for recurrent tidal flooding. This was accomplished through iterative MC error distribution modeling. An MC error distribution model was created to generate 100 unique permutations of the DEM, using a pseudo-random number generator and the bounds of potential error. The model follows the current practice of flood mapping which assumes that LiDAR vertical errors follow a normal distribution with zero bias (*28*). Similar methods were used by Liu et al. (*29*) for studying the effect of elevation error on shoreline position and Bodoque et al. (*30*) for characterizing first-floor elevation errors related to flood modeling.

For each increment of SLR, this study runs a tidal inundation model on each DEM permutation, recording differences between flood and elevation surfaces. The cumulative confidence of all inundation simulations was calculated by tallying the number of runs that resulted in each data cell being inundated. Grid cells were shaded in proportion to the number of simulations that produced flooding, providing a probabilistic delineation of potential error and flood vulnerability.

Thresholds and Tipping Points. For tidal flooding, we retained the NOAA threshold value for the study area based upon the local elevation of street-level impacts (*10*). Nonetheless, as tidal flooding increases frequency, today's nuisance or "extreme" becomes tomorrow's "mean" (*31*). The rate of occurrence of nuisance events is increasing along the East Coast, such that nuisance events are becoming chronic, and tipping points for impacts in areas such as Norfolk, where relative SLR rates are themselves faster than other areas.

Storm Surge Inundation Mapping

This research applies a multi-hazard methodology for vulnerability assessment. This includes applying the NOAA SLOSH model, superposition of SLOSH on morphostatic landform and developed surfaces with relative SLR, mapping surge inundation zones for three levels of storm severity, and estimating freeboard or potential depth of inundation of the critical structural elements. The research made use of existing and available data only and incorporated ancillary data characterizing the present-day risk conditions in the Federal Emergency Management Agency's updated Special Flood Hazard Areas published and provided in database form in mid-2016. Prior Virginia Department of Emergency Management flood zones are also used as reference to judge the conservativeness of our inundation models.

Inundation Model. For modeling hurricane surge inundation, we analyzed Saffir–Simpson categories 1–3 as surrogates for a range of hurricane surge severity. The Saffir–Simpson scale categorizes hurricanes into five classes which are determined the velocity sustained winds. Modeling of category 4 and 5 hurricanes was not necessary as storms of these intenstities produce surges which would completely inundate the study area and cause near-total destruction.

Inundation mapping implemented a hydrologically connected flood model, with existing baseline water at Mean Higher High Water (MHHW) datum expanding in eight possible directions (N, NE, E, SE, S, SW, W, NW) across the bare earth DEM for any cell less than or equal to the downscaled SLOSH surge height. Utilizing a cost-distance method, this produced a raster inundation grid. Each structural element was overlaid on the inundation grid for calculation of freeboard height (or depth in the case of complete inundation.) An inundation map was created for each category 1 through 3 for the baseline (today) as well as each future scenario of SLR. A common symbology color ramp and classification scheme was created to allow for comparison of the expanding inundation zone, with an underplayed orthophoto basemap image and superimposed point and polygon features of the structural assets.

Freeboard Elevation. Structural elements were characterized by a critical elevation value. These elevations were subtracted from the surge height grid, resulting in positive (above flood elevation) or negative (inundated) values. Thus, resulting tables characterize depth or freeboard and allow for ranking or classification of exposure. Tabular summaries were created by joining the measured freeboard-depth values to a master table containing the structural features' attributes. Tables were then symbolized and sorted by freeboard-depth for vulnerability analysis.

Tidal Flooding

Anecdotal reports of recurrent tidal flooding, often called "nuisance" flooding, in the Lafayette River basin and other areas of Norfolk have been increasing. Nuisance flooding is defined as a water level measured by NOAA tide gauges above the local NOAA National Weather Service threshold for minor impacts established for emergency preparedness (*11*). A 2016 Fugro (*32*) study provides support for these claims by indicating that tidal flooding in the Lafayette River area is frequent and is expected to worsen over time as "mean sea level" rises.

Well before 2050, it is expected that the number of days of tidal flooding will increase in pace with rising sea level. In 2016, Sweet and Marra (*11*) estimated the "nuisance flooding" level for Norfolk, VA to be 1.74ft (0.53 m) above MHHW datum and predicted an accelerating trend of tidal flooding days per year.

Two contrasting methods were employed for modeling of potential tidal flooding areas at NITS. First was a cost-distance model which simulated flooding only in land areas which are hydrologically connected to open water. Second was a bathtub model which uniformly simulates flooding in all land areas with elevations below the defined flood surface elevation, regardless of connectivity and barriers. The benefit of this composite methodology is that it allows for the identification of low-lying areas which are

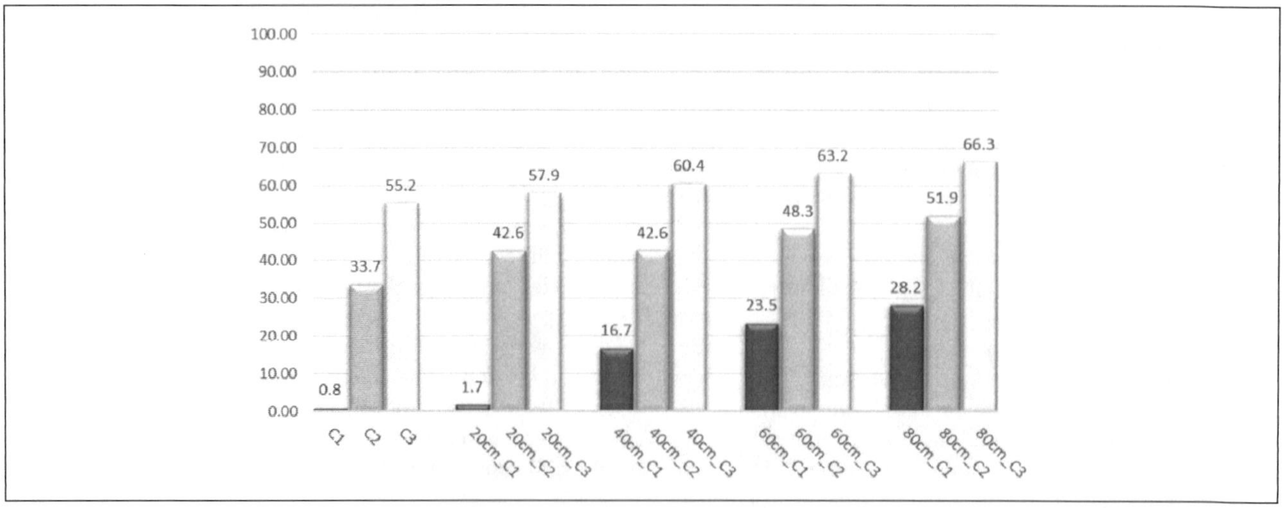

Figure 3. Percent of terminal potentially flooded by each surge for each sea level rise scenario. Increases in flooding percent between scenarios are expected to be non-linear as they are driven by elevation and topography of the study area.

not hydrologically connected to open water, but which are immediately adjacent and/or separated by narrow barriers. Each SLR scenario was modeled to include an additional 0.53 m of tidal inundation, the "nuisance" value calculated for Norfolk (*11*). Maps were created for comparison of the extents and both the cost-distance and bathtub-modeled flood surfaces.

Results

Storm Surge Inundation

The following sections summarize highlights and patterns of vulnerability for exposed assets at each sequential level of relative SLR. Maps of inundation patterns and tabular summaries of freeboard elevation (or depths) were analyzed iteratively and cumulatively with each SLR step.

Inundation Modeling and Mapping. Results for baseline surge at present day depict expected patterns of vulnerability extending along the outer fringe of the Lafayette River shoreline at category 1 level. Stronger storms (category 2–3) expose more extensive potential flooding around the pier and along reefer row and associated power substations. For moderate levels of surge (category 1–2), few critical elevations are attained by the surge MOMs. Although critical elevations are not attained, wave action, wind, and debris may nonetheless damage structures. In addition, a dozen assets may be inaccessible during peak surge (category 3). Percentage of ground area in the case study area predicted to be inundated by storms of identical category progressively increases as sea level increases (Figure 3). A gradient of progressive extent of inundation with category flood and increasing SLR is evident when mapped (Figure 4).

Tabulated Infrastructure Risk Exposure. Risk exposure summaries were tabulated for each item of critical infrastructure (Figure 5). The tables were grouped by infrastructure type and then sorted by elevation to reflect risk exposure from most to least (descending risk as elevation increases.) Color ranges from yellow, light orange, medium orange, orange-red and dark red symbolize the relative risk (from most to least) in direct proportion to the freeboard elevation from the surge water level up to the critical elevation of each element.

Results of the timing of potential exposure of each asset to surge inundation are complicated by uncertainty in future recurrence intervals of storm intensity. Nonetheless, assuming no stark increasing intensity, relative SLR will increase the risk of exposure to storm surges. With time, less surge can cause the same exposure, owing to the rise of static water levels. The first possible year of exposure for each structural element at the terminal was calculated by adding surge values to the NOAA high sea level curve value presented by the US Army Corps of Engineers Sea-Level Change Calculator, focusing on the level of critical elevation that a category 3 storm could impact.

Tidal Flooding

The current and future extent of recurrent "nuisance" tidal flooding in concert with relative SLR were also mapped (Figure 6). Results depict an expected pattern, with pronounced impacts on adjoining shorelines and residential areas of the Lafayette River, yet future tidal influence is principally confined by structures along the southern and southeast shore of the facility. At all modeled sea level scenarios, recurrent tidal flooding is expected to pose minimal risk to terminal infrastructure unless current barriers to tidal waters

Figure 4. Potential inundation maps for baseline sea level (today), +7.9" (20 cm) relative SLR, +15.7" (40 cm) relative SLR, +23.6" (60 cm) relative SLR, and +31.5" (80 cm) relative SLR scenarios.

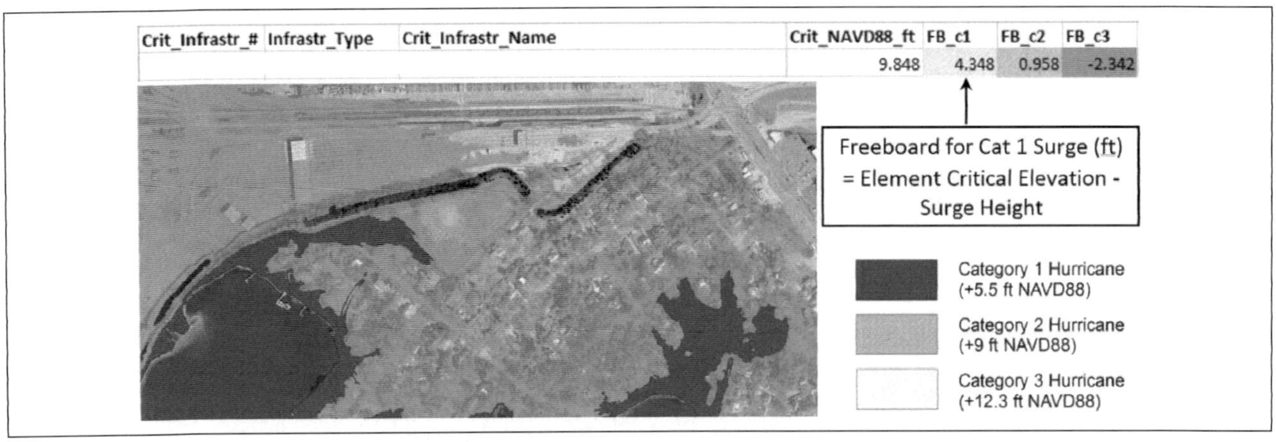

Figure 5. Example of infrastructure risk table usage. From left to right, columns FB_c1, FB_c2, and FB_c3 list the freeboard elevation in feet between the surge and the critical elevation of the structure. Positive values indicate critical elevation is not submerged and negative values indicate potential inundation depth. The tables also include the infrastructure type, name, and identification number, which is labeled in a series of inundation maps.

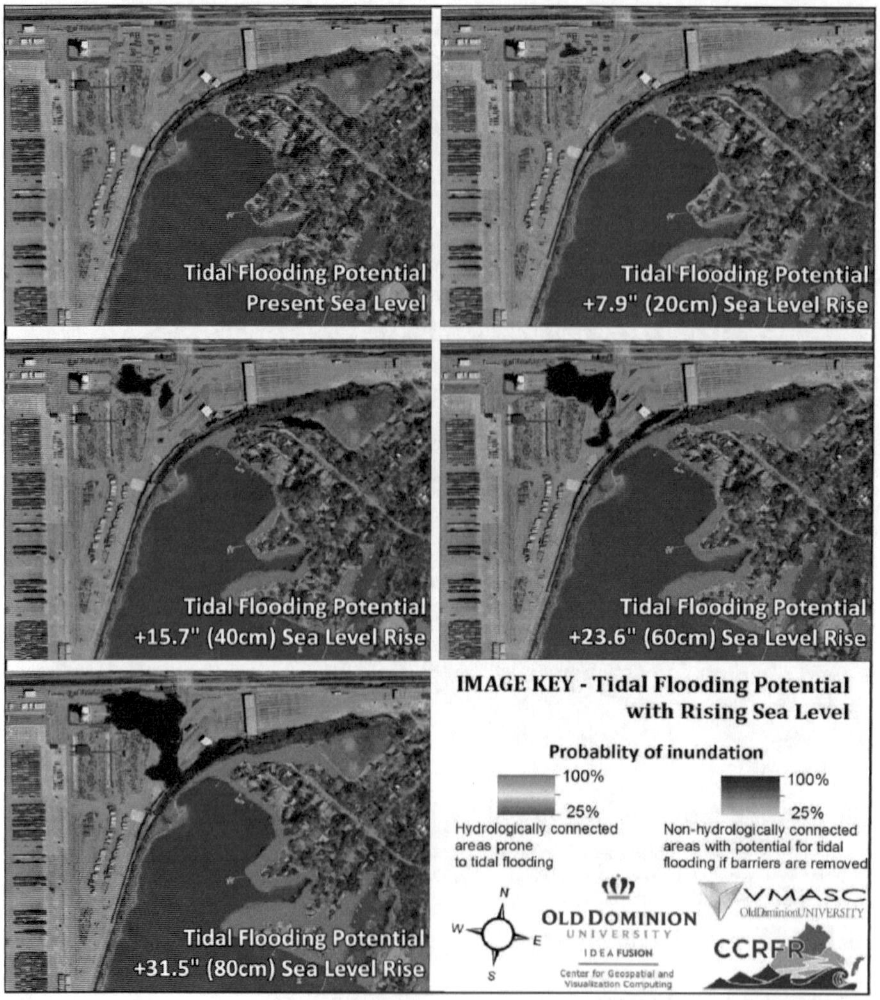

Figure 6. Tidal flooding simulations show the evolution of future tidal flooding from present (baseline) to possible future conditions with no human use modifications.

are removed or destroyed. Hardened structures that serve, increasingly, as levees during extreme tides could become susceptible to undermining and erosion or failure that could flood extensive disconnected low-lying areas.

Implications, Inferences, and Recommended Improvements

This risk assessment draws several significant inferences. First, identifying exposure of critical infrastructure is the first of several potential steps toward reducing the potential vulnerability of terminal assets. Second, the susceptibility of infrastructure is also an important ingredient to defining "vulnerability." Third, concise reporting of analysis results also precludes certain nuanced, specific differentiation of exposure of assets. For instance, use of a single critical elevation for a structure element may not capture exposure from submergence of adjoining terrain, which can prohibit access, emergency maintenance, or delay recovery. Cascading failures may also chain in sequences that were not considered in this assessment. Fourth, wave overtopping and velocity components, entrained debris, and wind may also cause damage in addition to surge submergence. For the longer time horizon, the relative elevation of ships and cargo may also bear consideration, as extreme tides today become mean tides in the future (*31*).

Strategic Actions to Improve Preparedness. This research recommends that the port operators engage in several strategic data development and research actions to improve preparedness: (1) ensure that vertical elevations of all infrastructure are accurately collected in both horizontal and vertical dimensions, (2) examine cascading failures in a systematic analysis, similar to the risk assessments of SERDP (Burkes-Copes) and Norfolk Naval Station, (3) investigate subsurface drainage and hydrologic connectivity, revealed here as ambiguous or incompletely determined; this analysis would also identify potential tailwater backup inhibiting rainfall runoff and causing ponding on tarmac and routine vehicle paths, and (4) collate data within an enterprise GIS, connecting this system to both planning, operations, and emergency management.

Planning Activities to Improve Preparedness. It is also recommended that the ports facing these risks engage in planning activities which will improve preparedness and extend the utility of this case study: (1) explore wider area vulnerabilities (e.g., transportation corridors adjoining the terminal), to include multi-modal transport linkages, in collaboration with surrounding municipal organizations, (2) incorporate continual monitoring or benchmarking of sea level and flood hazards into ongoing short-term planning, (3) incorporate vulnerabilities into plans for future disaster response and emergency management (including Hazmat waste and spills), (4) engage with regional entities on wider community risk assessment, adaptation, and mitigation projects and participate continually in these efforts, and (5) track the progress and status of other ports regionally, nationally, and internationally as they plan and adapt. This may be accomplished, in part, by the Port engaging with professional organizations, continually reviewing peer-reviewed and government research, and engaging in strategic business planning.

In recent years, significant progress was made by the HRSLR Preparedness and Resilience Intergovernmental Pilot Project to coordination communication activities (i.e., workshops, forums, town halls) which shared best practices in hazard mitigation and resilience between federal, state, local, and community organizations (*8*). Extending these efforts to focus on concerns specific to terminals and transportation corridors will be a critical step in improving port resilience.

Conclusions

The general conclusion is that marine terminal facilities should diligently plan for increased relative sea level and attendant increasing exposure to hurricane storm surges. Although there is marginal increased exposure of critical assets for smaller storm surges (using Saffir–Simpson category 1 and 2 as proxies) in the near-term, even moderate storm surges show potential to impact critical structures once relative sea level reaches +15.7–23.6" (20–60 cm). Given that those static levels of mean sea level could be attained as soon as the mid-to-late 2030s, prudence would dictate that major capital infrastructure investments ought to appropriately consider these findings.

An overall greater exposure is revealed by major hurricane surges (category 3 Saffir–Simpson scale as surrogate), and these lower probability events could nonetheless affect facilities sooner. As sea level rises, nearly all the infrastructure-critical elevation thresholds become vulnerable to a worst-case category 3 surge with +31.5" (80 cm) of relative SLR. Current precautionary (intermediate-high to extreme) sea level curves project this amount of rise could arise locally between the mid-2050s to early 2070s.

The approach used in this study does not evaluate future climate change insofar as it affects storminess, storm tracks, and hurricane frequency. In addition, extreme rainfall and changing climate affecting non-tropical storms and storm water flooding are not evaluated. Nonetheless, a range of implications are inferred and potential improvements undertaken to inform planning and operations. This preliminary assessment provides site- and element-specific data in map and tabular form. These data can be further leveraged if the Port were to develop a GIS database and integrate assets and vulnerabilities into enterprise planning and operations, along with regional observations (Sewell's Point tide gauge sea level trends) and regional data repositories (university, federal agencies, and city and regional government.)

In addition to storm surge and disaster preparedness, tidal flooding poses a creeping threat to this particular marine terminal through the nearby Lafayette River and potential shift of the MHHW tidal frame. As Sweet and Park (*31*) observe, today's extreme tides will become tomorrow's mean tides. The research's results point to undetermined subsurface hydro-connectivity between ponding on areas of the cargo tarmac and roads, the potential for "nuisance" tidal flooding today to increase in extent and frequency at moderate levels of SLR, and external considerations of street-level flooding as to transportation corridors surrounding this marine terminal.

Author Contributions

The authors confirm contribution to the paper as follows. Study conception and design: Thomas Allen, George McLeod, Joshua Behr; data collection: George McLeod, Thomas Allen; analysis and interpretation of results: Thomas Allen, George McLeod; draft manuscript preparation: George McLeod. All authors reviewed the results and approved the final version of the manuscript.

References

1. Burkes-Copes, K. A., E. J. Russo, Jr., , S. Bourne, M. Case, A. Davis, C. Fischenich, M. Follum, H. Li, L. Lin, S. Lofton, K. McKay, P. Mlakar, A. Morang, S. Pranger, R. Pickett, J. Ratclif, J. Rullán-Rodríguez, M. Schultz, J. Sims, E. Smith, J. Smith, C. Talbot, and K. Winters. *Risk Quantification for Sustaining Coastal Military Installation Asset and Mission Capabilities (RC-1701)*, Final Report, The Strategic Environmental Research and Development Program (SERDP), U.S. Army Engineer Research and Development Center (ERDC), Vicksburg, MS, 2008.
2. Titus, J.G., and D. Cacela. Uncertainty Ranges Associated with EPA's Estimates of the Area of Land Close to Sea Level. In

Background Documents Supporting Climate Change Science Program Synthesis and Assessment Product 4.1: Coastal Elevations and Sensitivity to Sea Level Rise (J. G. Titus and E. M. Strange, eds.), U.S. EPA, Washington, D.C., 2008.

3. North Carolina Department of Emergency Management. *North Carolina Sea Level Rise Risk Management Study: Conceptual Model Report*. NC Division of Emergency Management, Office of Geospatial and Technology Management. V.2.2, 2011. https://ericnatthomas.files.wordpress.com/2013/12/conceptual-model-report_ver2-2_20110502.pdf
4. Van Houtven, G., B. Depro, D. Lapidus, J. Allpress, and B. Lord. *The Cost of Doing Nothing: A Sea Level Rise Synopsis for the Hampton Roads Region*. Virginia Coastal Policy Clinic RTI International, Washington, D.C., 2016.
5. Belfield, S. S., and B. McFarlane. *Sea Level Rise and Storm Surge Impacts to Roadways in Hampton Roads*. Hampton Roads Transportation Planning Organization, Chesapeake, VA, 2016.
6. *Tidal Inundation Tracking Application for Norfolk (TITAN)*. City of Norfolk, GIS Team, Department of Communications and Technology. http://gisapp1.norfolk.gov/Titan/. Accessed June 14, 2017.
7. Loftis, J. D., H. V. Wang, R. J. DeYoung, and W. B. Ball. Using Lidar Elevation Data to Develop a Topobathymetric Digital Elevation Model for Sub-Grid Inundation Modeling at Langley Research Center. *Journal of Coastal Research*, 2016, pp. 134–148. 2016.
8. Steinhilber, E. E., M. Boswell, C. Considine, and L. Mast. Hampton Roads Sea Level Rise Preparedness and Resilience Intergovernmental Pilot Project Phase 2 Report: Recommendations, Accomplishments and Lessons Learned. *Hampton Roads Intergovernmental Pilot Project: Reports. 2*. https://digitalcommons.odu.edu/hripp_reports/2/. Accessed June 13, 2017.
9. Hall, J. A., S. Gill, J. Obeysekera, W. Sweet, K. Knuuti, and J. Marburger. *Regional Sea Level Scenarios for Coastal Risk Management: Managing the Uncertainty of Future Sea Level Change and Extreme Water Levels for Department of Defense Coastal Sites Worldwide*. U.S. Department of Defense, Strategic Environmental Research and Development Program, Washington, D.C., 2016, 224 pp.
10. Sweet, W. V., R. E. Kopp, C. P. Weaver, J. Obeysekera, R. M. Horton, E. R. Thieler, and C. Zervas. *Global and Regional Sea Level Rise Scenarios for the United States*. NOAA Technical Report NOS CO-OPS 083, US Department of Commerce, 2017, 56 pp. https://tidesandcurrents.noaa.gov/publications/techrpt83_Global_and_Regional_SLR_Scenarios_for_the_US_final.pdf.
11. Sweet, W., and J. Marra. *State of U.S. "Nuisance" Tidal Flooding*. National Oceanic and Atmospheric Administration's Center for Operational Oceanographic Products and Services and National Centers for Environmental Information, Silver Spring, MD, https://www.ncdc.noaa.gov/monitoring-content/sotc/national/2016/may/sweet-marra-nuisance-flooding-2015.pdf. Accessed June 8, 2016. 5 pp.
12. Akukwe, T. I., and C. Ogbodo. Spatial Analysis of Vulnerability to Flooding in Port Harcourt Metropolis, Nigeria. *SAGE Open*, Volume 5 Issue 1, 2015, pp. 1–19. https://doi.org/10.1177/2158244015575558.
13. Hallegatte, S., N. Ranger, O. Mestre, P. Dumas, J. Corfee-Morlot, C. Herweijer, and R. Muir Wood. Assessing Climate Change Impacts, Sea Level Rise and Storm Surge Risk in Port Cities: A Case Study on Copenhagen. *Climatic Change*, Vol. 104, 2011, pp. 113–137.
14. Becker, A., S. Inoue, M. Fischer, and B. Schwegler. Climate Change Impacts on International Seaports: Knowledge, Perceptions, and Planning Efforts Among Port Administrators. *Climatic Change*, Vol. 110, 2012, pp. 5–29.
15. Hanson, S., R. Nicholls, N. Ranger, S. Hallegatte, J. Corfee-Morlot, C. Herweijer, and J. Chateau. A Global Ranking of Port Cities with High Exposure to Climate Extremes. *Climatic Change*, Vol. 104, 2011, pp. 89–111.
16. Nicholls, R. J. *Ranking Port Cities with High Exposure and Vulnerability to Climate Extremes: Exposure Estimates*. OECD Environment Working Papers, No. 1, OECD Publishing, Paris, 2008. http://www.oecd.org/dataoecd/16/58/39720578.pdf
17. Rajabalinejad, M., and Z. Demibilek. A Bayesian Probabilistic Approach for Impacts of Sea Level Rise on Coastal Engineering Design Practice. *Ocean Engineering*, Vol. 71, 2013, pp. 66–73.
18. Sierra, J. P., I. Casanovas, M. Mosso, M. Mestres, and A. Sanchez-Arcilla. Vulnerability of Catalan (NW Mediterranean) Ports to Wave Overtopping Due to Different Scenarios of Sea Level Rise. *Regional Environmental Change*, Vol. 16, 2016, pp. 1457–1468.
19. Hinkel, J., C. Jaeger, R. J. Nicholls, J. Lowe, O. Renn, and S. Peijun. Sea-level Risk Scenarios and Coastal Risk Management. *Nature Climate Change*, Vol. 5, 2015, pp. 188–190.
20. Good, J. W., R. F. Goodwin, and N. J. Wood. Reducing Vulnerability of Ports and Harbors to Earthquake-tsunami Events. Converging Currents: Science, Culture, and Policy at the Coast. *Proc., 18th International Conference of the Coastal Society*, Galveston, TX USA. National Sea Grant Library. 2001.
21. National Hurricane Center. *SLOSH Model*. Retrieved from Sea, Lake, and Overland Surges from Hurricanes. http://www.nhc.noaa.gov/surge/slosh.php. Accessed March 24, 2015.
22. Allen, T. R., S. Sanchagrin, and G. McLeod. Visualization for Hurricane Storm Surge Risk for Awareness and Emergency Communication. In *Approaches to Disaster Management – Examining the Implications of Hazards, Emergencies and Disasters* (J. Tiefenbacher ed.), InTech Publishing, 2013. https://www.intechopen.com/books/approaches-to-disaster-management-examining-the-implications-of-hazards-emergencies-and-disasters/visualization-for-hurricane-storm-surge-risk-awareness-and-emergency-communication
23. Huber, M., and K. White. *Sea Level Change Curve Calculator (2015.46) User Manual*. US Army Corps of Engineers. 2015, 21 pp. http://www.corpsclimate.us/docs/Sea_Level_Change_Curve_Calculator_User_Manual_2015_46_FINAL.pdf
24. Gesch, D. B. Elevation Uncertainty in Coastal Inundation Hazard Assessments. In *Natural Disasters* (S. Cheval, ed.), InTech, Rijeka, Croatia, 2012, pp. 121–140.
25. Gesch, D. B. Analysis of Lidar Elevation Data for Improved Identification and Delineation of Lands Vulnerable to Sea-Level Rise. *Journal of Coastal Research*, Vol. 53, 2009, pp. 49–58.

26. Heidemann, H. K. *Lidar Base Specification (ver. 1.2)*. U.S. Geological Survey Techniques and Methods, Reston, Virginia, 2014, 11-B4, 67 pp.
27. Dewberry & Davis LLC. *USGS Norfolk, VA LiDAR*. USGS Contract: G10PC00013, Task Order: G3PD00279. U.S. Geological Survey, 2014, 233 pp. https://ftp.vgingis.com/Download_2/LiDAR/2013/norfolk/Project_Report_Norfolk_VA_LiDAR_20140530.pdf
28. Cooper, H., C. Fletcher, Q. Chen, and M. Barbee. Sea-level Rise Vulnerability Mapping for Adaptation Decisions Using LiDAR DEMs. *Progress in Physical Geography*, Vol. 37, No. 6, 2013, pp. 745–766.
29. Liu, H., D. Sherman, and S. Gu. Automated Extraction of Shorelines from Airborne Light Detection and Ranging Data and Accuracy Assessment Based on Monte Carlo Simulation. *Journal of Coastal Research*, Vol. 23, No. 6, 2007, pp. 1359–1369.
30. Bodoque, J., C. Guardiola-Albert, E. Aroca-Jimenez, M. Eguibar, and M. Martinez-Chenoll. Flood Damage Analysis: First Floor Elevation Uncertainty Resulting from LiDAR-Derived Digital Surface Models. *Remote Sensing*, Vol. 8, 2016, pp. 604.
31. Sweet, W. V., and J. Park. From the Extreme to the Mean: Acceleration and Tipping Points of Coastal Inundation from Sea Level Rise. *Earth's Future*, Vol. 2, No. 12, 2014, pp. 579–600.
32. Fugro Consultants, Inc. *Lafayette River Tidal Protection Alternatives Evaluation*. Work Order No. 7. Fugro Project No. 04.8113009. City of Norfolk, City-wide Coastal Flooding Contract, 2016, 56 pp. https://www.norfolk.gov/DocumentCenter/View/25170

The Standing Committee on Ports and Channels (AW010) peer-reviewed this paper (18-02516).

Article

TRR
JOURNAL OF THE TRANSPORTATION RESEARCH BOARD

Transportation Research Record
2018, Vol. 2672(11) 30–40

DOI: 10.1177/0361198118796353
journals.sagepub.com/home/trr
SAGE

Developing and Implementing a Port Fluidity Performance Measurement Methodology using Automatic Identification System Data

C. James Kruse[1], Kenneth N. Mitchell[2], Patricia K. DiJoseph[2], Dong Hun Kang[3], David L. Schrank[3], and William L. Eisele[3]

Abstract

The U.S. Army Corps of Engineers (USACE) is responsible for the maintenance of federally authorized navigation channels and associated infrastructure. As such, USACE requires objective performance measures for determining the level of service being provided by the hundreds of maintained navigation projects nationwide. To this end, the U.S. Army Engineer Research and Development Center partnered with Texas A&M Transportation Institute to develop a freight fluidity assessment framework for coastal ports. The goal was to use archival automatic identification system (AIS) data to develop and demonstrate how ports can be objectively compared in relation to fluidity, or the turnaround time reliability of oceangoing vessels. The framework allows USACE to evaluate maintained navigation project conditions alongside port system performance indices, thereby providing insight into questions of required maintained channel dimensions. The freight fluidity concept focuses on supply chain performance measures such as travel time reliability and end-to-end shipping costs. Although there are numerous research efforts underway to implement freight fluidity, this is the first known application to U.S. ports. This paper covers AIS data inputs, quality control, and performance measures development, and also provides a demonstration application of the methodology at the Port of Mobile, Alabama, highlighting travel time and travel time reliability operating statistics for the overall port area. This work provides foundational knowledge to practitioners and port stakeholders looking to improve supply chain performance and is also valuable for researchers interested in the development and application of multimodal freight fluidity performance measures.

What is Freight Fluidity?

The concept of a "fluidity indicator" was first popularized by Transport Canada to evaluate the performance of trade corridors and multimodal supply chains. For Transport Canada's applications, the fluidity indicator measures total transit time and travel time reliability of goods along defined supply chains (*1*). Texas A&M Transportation Institute provided early technical assistance to Transport Canada to develop and test fluidity measures and demonstrate them along a supply chain, which is documented elsewhere (*2*).

Mr Louis-Paul Tardif of Transport Canada—who is credited with coining the phrase freight fluidity—stated that "the fluidity indicator provides evidence-based information to assess and analyze the efficiency of supply chains and assists Transport Canada's work in identifying constraints in the transportation system" (*3*).

Freight fluidity refers to the performance of transportation supply chains and freight networks. It can be a measure of the performance of a supply chain using a single mode or multiple modes of freight transportation. Freight fluidity can also be a measure of the performance of a freight network or freight corridor serving many supply chains.

In current U.S. parlance, freight fluidity focuses on transportation supply chain performance measurement (SCPM) (*4*), that is, the measurement of travel time, travel time reliability, and cost of moving freight shipments from end to end of a supply chain. For this paper,

[1]Center for Ports & Waterways, Texas A&M Transportation Institute, The Texas A&M University System, College Station, TX
[2]US Army Engineer Research and Development Center Coastal and Hydraulics Laboratory, Vicksburg, MS
[3]Texas A&M Transportation Institute, The Texas A&M University System, College Station, TX

Corresponding Author:
Address correspondence to C. James Kruse: j-kruse@tti.tamu.edu

the maritime portion of the supply chain, specifically the time required for vessels to call at deep-draft coastal ports, is the focus.

What Does Freight Fluidity Mean for Ports?

The transportation network (and supply chain network) is multimodal and complex. Ports are a very important link in global supply chains. To maintain consistency with other freight fluidity efforts at regional, national, and international levels, the authors recommend that the definition of port fluidity preserve the focus on transportation SCPM (*4*)—primarily the measurement of travel time, travel time reliability, and cost of moving freight shipments from end to end of a supply chain.

Freight fluidity for ports focuses on travel times, dwell times (at terminals or anchorage locations), and travel time reliability. It should be noted that automatic identification system (AIS) data only capture the movement of individual marine vessels and do not provide direct insight into the volumes of cargo loaded or unloaded during any particular terminal visit. Nonetheless, it is the authors' contention that AIS provides very useful proxy data that helps provide quantitative insights into port system performance, specifically fluidity trends through time. The costs (whether actual or opportunity costs) are less straightforward to quantify for vessels. Unlike for roadway users (trucks), where wasted time and fuel in congestion can be readily computed, these cost factors are more difficult to obtain for vessels. Tackling the issue of costs is recommended after a clear understanding of travel time and travel time reliability are developed. One approach is to initially identify potential tonnage affected by delays as a surrogate for costs.

Background and Need

USACE is responsible for the maintenance of federally authorized navigation channels, both in deep-draft coastal ports as well as along the inland river system. In recent decades, annual funding outlays for maintenance dredging have generally not kept pace with the continuing need to remove accumulating sediment from the nation's ports, harbors, and access channels. Much attention has been given to this matter by a number of industry associations (e.g., the American Association of Port Authorities, regional port associations, and others) as well as individual port authorities and their users. Though a lack of maintenance dredging clearly has a detrimental effect, in a limited funding environment USACE must take many aspects of navigation project condition into account and evaluate the maintenance dredging needs at hundreds of projects across the country. In so doing, USACE needs objective, quantitative measures to help determine whether it is adequately meeting its navigation mission to "provide a safe, reliable, efficient, effective, and environmentally sustainable waterborne transportation system for movement of commerce, national security needs, and recreation" (*5*, *6*). Stated differently, one of USACE's primary responsibilities is to ensure that federal navigation projects are adequately maintained for the particular vessels that actually use them.

This is obviously of greatest importance where frequent channel maintenance dredging is required. There are indications that the resources provided to USACE are not adequate for keeping the channels at authorized dimensions. For example, in the USACE Civil Works Program Five-Year Development Plan: Fiscal Year 2011 to Fiscal Year 2015, USACE reported that for fiscal year 2005 to fiscal year 2008 (*7*), the center half of the most-used deep-draft navigation channels was only available at authorized dimensions between 30% and 38% of the time. This is based on values for the top 59 coastal and Great Lakes navigation projects based on tonnage; all projects included handled more than 10 million tons.

With the advent of AIS technology for maritime domain awareness, vessel operators have an unprecedented capability to "see and be seen," thereby significantly enhancing safety and security for mariners and public alike. AIS transceiver units have been mandated for most commercial vessels since 2003 (33 CFR § 164). The primary need and uses of AIS for ship-to-ship communication and safety, such as its use as a real-time data feed, is defined by the International Telecommunication Union (*8*). In recent years, several authors have looked into the use of archival AIS data for waterway management and performance monitoring (*9–13*), these and similar efforts touch on various important issues that pertain to this paper.

- The Coast Guard is responsible for recording and archiving all AIS transmissions that pertain to the United States, and it does so through its nationwide automatic identification system (NAIS). USACE also provides the Coast Guard with the AIS transmissions it collects (*14*).
- Over time, the Coast Guard has attempted to improve the quality of the AIS dataset as and when technical issues arise with its use (*15*).
- In very recent years, use of the AIS system has expanded into the inland waterways throughout the United States. (*15*).
- The availability and use of AIS data at a localized level has been investigated by many authors (*16*).
- AIS data have shown patterns in vessel behavior within some broad vessel classes (*16*).

- AIS vessel inventory information such as dimensions (e.g., length, draft) require additional scrutiny by users (*16*).

Objective

The primary objective of this paper is to use archival AIS data to develop and demonstrate freight fluidity measures that will allow USACE to evaluate federal navigation projects in relation to their suitability for the vessel types and classes that presently use them. These measures will focus on the summary statistics from three essential performance attributes of navigation channels, allowing USACE to evaluate and compare the conditions of its channels and answer the following questions.

1. What is the transit time for vessels from the anchorage area to the terminals and vice versa?
2. What is the dwell time at the defined nodes?
3. How do transit restrictions (e.g., one-way or daylight restrictions) affect transit or dwell times?
4. Have there been any changes in these performance attributes year-to-year or before and after an event?

Analysis Overview

There are several important steps to work through to develop and test a methodology at a U.S. port. These steps are important to understand the port being analyzed, the data used, and the measures in the analysis. These steps are outlined below and will be expanded upon in this paper.

- Port selection: It is important to select a port that has a diverse cargo menu but is not too large (in terms of vessel traffic) to ensure testing is not overly complex. The authors chose the Port of Mobile as a suitable test bed.
- Input data: The key data used in the mobility analysis.
- Input parameters: Some parameters need to be defined as they relate to the port facilities. These define the anchorage area, dock locations, the channel to traverse the port, and the trip components associated with a port visit.
- Performance measure definitions: The performance measures need to be defined for waterway travel and data, which have slightly different considerations than roadway travel and data.

The authors contend that upon successful completion of the methodology testing at the Port of Mobile (discussed in the remaining sections), the methods should be adaptable for other port locations.

Port of Mobile Methodology Implementation

The authors conducted an analysis on the Port of Mobile, Alabama to develop and implement a marine freight fluidity performance management methodology. Mobile has several characteristics that make it an excellent location to demonstrate and implement the methods:

- Location: It is on the Gulf of Mexico. The continental shelf in the Gulf of Mexico is shallow and has a very low gradient. Furthermore, the ocean bed consists of soft material that is highly fluid and necessitates frequent dredging.
- Cargo Mix: It has a fairly comprehensive mix of cargo and vessel types. In 2014, 36.8 million tons of foreign cargo were handled in Mobile Harbor. This total can be broken down into major commodity categories as follows (*17*):
  - Coal: 15.9 million tons
  - Petroleum and petroleum products: 5.7 million tons
  - Crude materials: 4.8 million tons
  - Manufactured goods: 6.6 million tons
  - All others: 3.8 million tons.

The movement of this cargo involved the transfer of 238,000 twenty-foot-equivalent units of container traffic (*18*).

- Size: It is a mid-sized port based on vessel traffic and is therefore manageable in terms of the effort needed to develop a methodology and implement a pilot program, yet it is large enough that resulting techniques should be generalizable to other U.S. ports to expand the methods. In 2014, Mobile ranked 10th in the nation in terms of tonnage handled; however, it was 6th among Gulf of Mexico seaports (*19*). It was 21st nationally in terms of vessel calls, according to U.S. Maritime Administration statistics; again, it was 6th among Gulf ports (*20*).

Input Data: AIS

Some of the key data attributes and descriptions of these attributes from the AIS dataset are shown in the list below (*9*). These data were obtained from the U.S. Coast Guard's NAIS archive of vessel position reports using a suite of web services provided to USACE as part of a data-sharing interagency security agreement. These key variables include the ship identification numbers (maritime mobile service identity [MMSI] and IMO_NUMBER), the physical georeferenced location (LAT, LON), and the date/time stamp (TX_DTTM). Other variables that were used in the analysis included NAME and SHIP_AND_CARGO_TYPE, which shows the type of vessel and type of cargo. The MMSI and IMO_NUMBER were the key variables used to identify each individual vessel whereas the LAT/LON and TX_DTTM variables allowed a determination of the location of the vessel at any given time to track movement.

- MMSI: vessel unique ID;
- IMO_NUMBER: international maritime organization unique ID;
- TX_DTTM: date-time stamp;
- LAT: latitude coordinate in decimal degrees;
- LON: longitude coordinate in decimal degrees;
- NAME: vessel name; and
- SHIP_AND_CARGO_TYPE: type of vessel and cargo type.

Input Parameters Definition

Some basic parameters are needed to analyze port mobility. It is important to note that there are two sub-port complexes at the Port of Mobile. The Theodore facility is on the western side of Mobile Bay and is at about the halfway point in the channel that traverses the bay. The larger facility is referred to as the River Harbor facility in this analysis, and it encompasses all of the dock facilities at or beyond the entrance to Mobile River at the north end of the bay. Specific to the Port of Mobile, the trip types were defined as follows for the port movements and dwell times, which are shown in Figure 1:

- Total port system time (TPST): the total elapsed time from the moment the vessel approaches the outer anchorage area until it exits the port. The outer anchorage area is the polygon (box) area; once a vessel enters this area, it is included in the analysis. The TPST represents the time that accrues between #1 and #5 (Figure 1).
- Port cycle time (PCT): the total elapsed time from the moment the vessel enters the port until it crosses the same point to exit. This represents the time that accrues between #2 and #5 (Figure 1).
- Travel time inbound (TTIB): the travel time from the point the vessel enters the channel until it nears the dock area. This represents the time that accrues between #2 and #3 (Figure 1).
- Travel time outbound (TTOB): the travel time from the point the vessel leaves the dock area, travels through the channel and crosses back into the "catch" area on its way out. This represents the time that accrues between #4 and #5 (Figure 1).

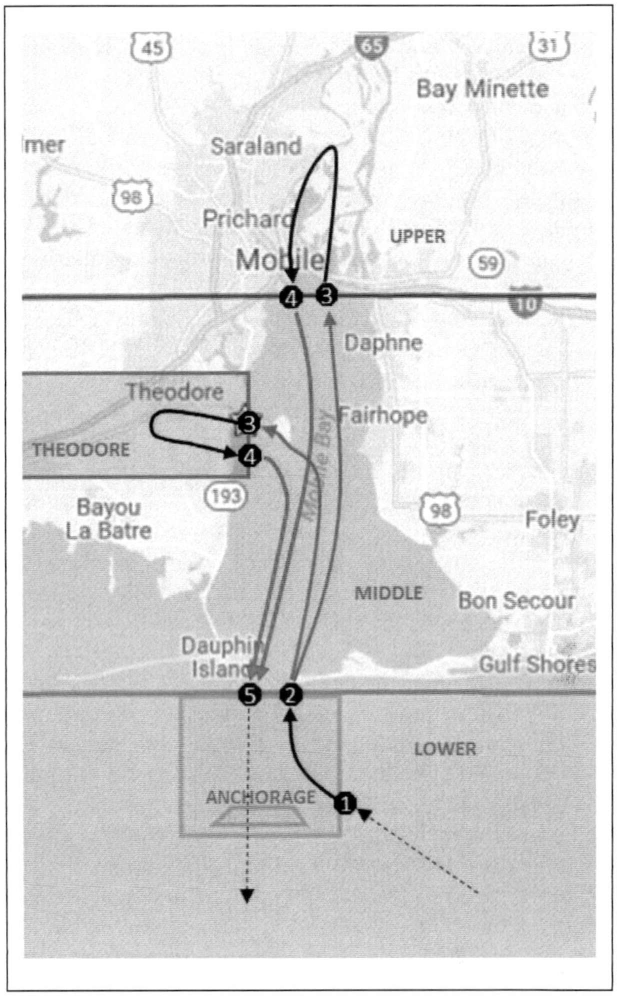

Figure 1. Map of the Port of Mobile activities.

Performance Measures

The next step of the port freight fluidity methodology is the determination of the performance measures to include and what parameters will be included within these parameters.

The first measure that is needed is a baseline travel time for the comparison for each trip that is made through the port. This measure identifies a "freeflow" or minimum amount of travel time necessary to complete the trip—in other words, an unobstructed transit. On roadways, this baseline condition is often called a free-flow travel time, and it is defined as the travel times that are observed during light traffic conditions. Physical operating conditions are different on the water (and in a port) therefore different baseline logic needs to be employed. The researchers reviewed two alternatives: the 15th and 25th percentile travel times. On roadways, the 15th percentile travel time is often associated with free-flow conditions (i.e., 85th percentile speed). USACE staff have used the 25th percentile travel time for baseline conditions in other analyses therefore the researchers included the 25th percentile travel time in the exploratory analysis to determine an appropriate baseline percentile.

After analyzing the results using both the 15th and 25th percentiles, the researchers chose the 25th percentile because it is less sensitive to long tails (i.e., high travel times) in the trip distribution, as evidenced by visual inspection of the resulting performance measures. Based on researcher experiences, when inappropriate baseline travel times are used as percentiles, performance measures can vary substantially and may appear inflated in some cases. The results of the percentile investigation are included in the project report but are not shown in this paper due to length constraints (*21*).

Similarly, the median and mean travel times were both analyzed to determine which should be used to represent the "average" condition. In this analysis, median travel time was chosen because it too is less sensitive to high travel times.

The travel time index (TTI) is the average of the ratios of each individual trip travel time to the baseline travel time (25th percentile) for the set of these individual trips. This shows mobility conditions for each trip relative to the baseline (25th percentile) travel time within the port facilities.

The planning time index (PTI) is the ratio of the 95th percentile travel time of a set of trips to the baseline travel time for the set of trips. A typical mobility analysis is a review of the variability of travel times—referred to as travel time reliability. The researchers used the 95th percentile travel time to analyze travel time reliability in the port because some cargo requires an on-time delivery and there are penalties for missing a predetermined delivery period. Therefore, using the 95th percentile travel time ensures that the vessel will arrive at the dock within a more stringent and designated time window (*22*).

Time above a baseline travel time is the extra travel time needed to complete the trip. In mobility analyses on roadways this would be considered "travel delay." However, with port operations not all of this additional time is truly "delay" as some of the additional time could be due to things such as additional moored time while waiting for the next port call to be assigned. Thus, this additional time above the baseline is simply referred to as "time above the baseline travel time."

Initial Raw AIS Data Exploration and Preparation

USACE provided AIS data for calendar years 2014 and 2015 for all vessels transiting within the area (Figure 1). The first step in the analysis of this data included becoming familiar with the AIS database to ultimately determine valid data for the final analysis. Some preliminary screening was performed on the AIS database using the geographic location along with vessel type, name, and valid MMSI and International Maritime Organization (IMO) numbers. Vessels identified as noncargo by the AIS SHIP_AND_CARGO_TYPE inventory information were eliminated. Vessels without inventory information were also eliminated in the initial screening. All of these actions resulted in a filtered dataset consisting of approximately 9 million vessel position reports and associated time stamps.

After the initial filtering, the vessel trip data were further reviewed geospatially along the paths so that the vessel data points associated with key trip points labeled as 1 through 5 were identified (Figure 1). Each LAT/LON of a vessel that passed each of the X Y coordinates assigned to trip points 1 through 5 were used to identify when that vessel crossed that key trip point (Figure 1). Any intermediate data points between any two of the LAT/LON data associated with the trip points were eliminated because they were not needed. Only the time stamps associated with each of the key trip points 1 through 5 for each vessel were needed to track the movements. Next, the trips with any missing trip points were excluded therefore the performance measures were calculated based on the trips with all five of the data points. A vessel could be missing a trip point because the AIS system failed to report for a certain period of time or because of obstructions from weather, terrain, or both weather and terrain. This occurred infrequently. All of the filtering actions to this point reduced the data to about 14,000 records.

Upon investigating the distributions of the various vessel travel times, some very long durations were observed in the dataset. This observation encouraged caps to be placed on trip times to eliminate trips that appeared to be extremely long and possibly involved other activities in addition to normal cargo operations (e.g., drydocking). The following caps were used:

- TPST: 2 weeks;
- PCT: 1 week; and
- Transit time inbound or outbound: twice the median travel time (or about 5 h).

These additional actions, based on valid travel times, lowered the data size from about 14,000 to about 12,000 records. This very simple screening effort revealed that

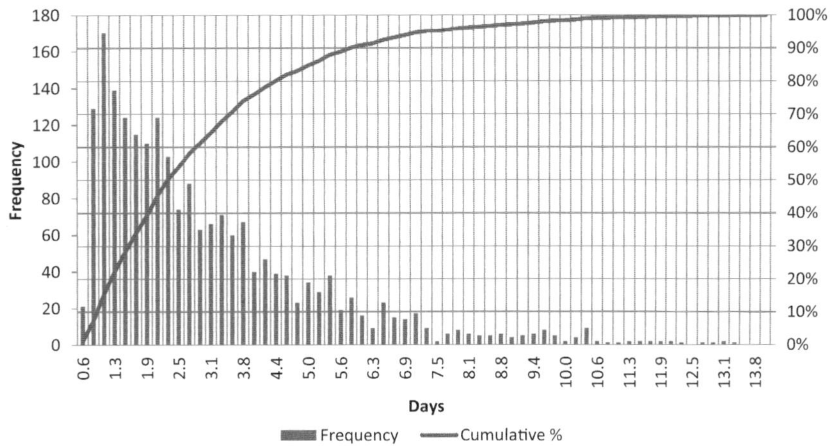

Figure 2. Histogram of total TPST.

using the AIS vessel inventory by itself could be problematic due to errors in the operator-input AIS message fields. This was made evident by analyzing the remaining trips and travel times that were still included in the data. Some of these seemed odd for cargo vessels, causing the researchers to dig further into the data. The initial screening caused many valid vessels to be eliminated and many invalid vessels to be retained due to incorrect or vague coding of the inventory data. A second, more detailed screening of the AIS data using Lloyd's Register inventory data, described later in this document, revealed a very different dataset of vessels and indicated that the initial screening of the dataset using only AIS inventory information would not be sufficient.

The distribution of travel times for the TPST and the average time spent in the anchorage area at the entrance to the port following the initial screening of the data are shown (figures 2 and 3). These graphics show the results of the caps that were placed on trip times for the various trip types. The distribution of travel times for the TPST are shown (Figure 2). There are quite a few trips (around 2%, or 38 trips) that have total TPST times of 10 days or more, meaning that 2% of calls on the Port of Mobile went over 10 days. The amount of time spent in the anchorage area is shown in Figure 3. This graphic was cut off at about 70% of trips because the long tail of anchorage time continued well beyond the 8 h shown as many vessels had to anchor for a day or more while awaiting entry into the port. Approximately 65% of the trips spent 4 h or less in the anchorage area prior to entering the port, therefore a vast majority of trips spend very little time awaiting entry into the port. About 26% of trips spend at least 10 h or more in the anchorage area.

Similar to the analysis of the mean and median for travel times, the 15th and 25th percentile travel times were calculated to determine a freeflow or baseline travel time to use in determining which vessel calls included extra time in Mobile.

Initial exploration of the baseline travel time showed very similar results between the inbound and outbound transit times, so only the inbound transit time is shown in the analysis. The inbound transit time often has a higher urgency of the two trips using the port channel because the vessel is loaded and has a delivery window to meet, although the outbound trip can be equally urgent if the vessel must meet a deadline at another port.

As previously discussed, the median travel time was used rather than the mean travel time to capture representative conditions in this analysis. The distribution of TPST for all vessel types at River Harbor is shown (Figure 4). River Harbor is shown because all vessel types docked there, whereas Theodore did not have container or crude oil tanker vessels. Some vessel types, namely dry bulk cargo vessels, have much longer tails in their travel time distributions, with TPST trip lengths of 10 to 14 days, whereas others, such as the container vessels, tend to have very short stays in port, which tend to have a TPST around 2 to 3 days (Figure 4). These very different distributions caused the mean travel times to be much greater than the median travel times for some of these vessel type populations. Due to the longer tails for some cargo types, the mean and median values are separated by about 15 h when all trips are grouped together.

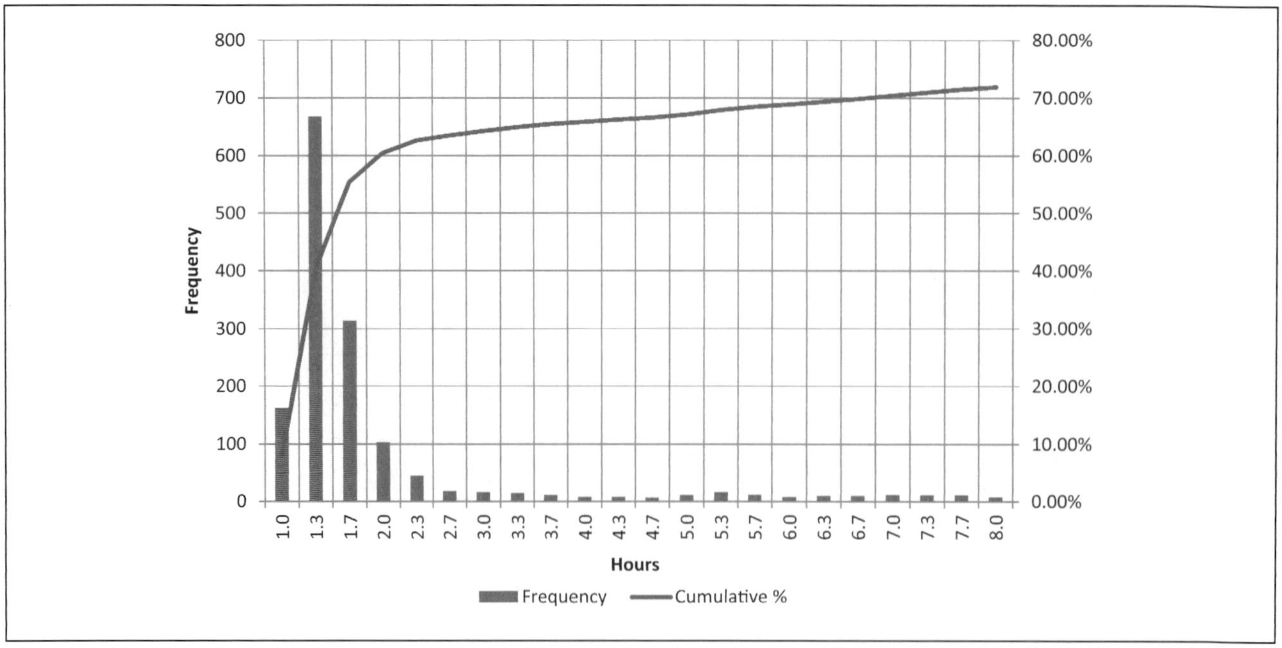

Figure 3. Histogram of mobile anchorage area dwell times.

To calculate mobility performance measures, the baseline travel time is required for comparison purposes for each vessel trip. This baseline value is calculated for each individual analysis being performed. If all of the vessels are included in the analysis, then all vessel trips are used to calculate the baseline travel time. If a subset of vessels is being analyzed (e.g., container vessels) then the baseline travel time is calculated for just those vessels (in this case, container vessels). This can cause some of the tabular results to look unintuitive as the overall summary for the data may not fall within the range of those numbers due to different baselines for each row in a table.

Supplemental Datasets Used and Secondary Data Preparation

Further cleaning of the AIS dataset involved identifying all MMSI/IMO numbers in the dataset that appeared in the 2-year period at the Port of Mobile and using the Lloyd's Register database to query pertinent information such as name, ship type, size, and so forth. This information was used to determine the vessels that were to be included in the final analysis shown in the next section. Non-cargo-carrying vessels such as tugs, dredges, other service vessels, pleasure craft, fishing vessels, and so on were eliminated using the Lloyd's data; some of these vessels had missing or problematic data in the original AIS vessel characteristics data.

Analysis Results

Overall Mobility Levels

Following the second AIS database cleanup to eliminate non-relevant vessels, there were 2,165 trips analyzed from 2014 and 2015. The majority of these trips (1,997) were associated with River Harbor whereas only 168 were at the Theodore facility. The large discrepancy in the number of vessels is primarily due to the types of vessels served at each facility and the overall size of each facility. The freeflow or baseline travel times (25th percentile travel time) for the various trip types are shown for both facilities (Table 1). The baseline for TPST is 33.2 h for all vessels entering the Port of Mobile, meaning that for 25% of the port calls it takes just over 33 h for a vessel to complete its entire trip into and out of the port. The baseline for PCT is 28 h. This means that, on average, vessels are in the anchorage area for approximately 5 h, which is the difference between the TPST and PCT. Vessels that steam straight into the port without much wait time in the anchorage area are included in these calculations with the vessels that drop anchor.

The overall TTI for TPST is 1.74, which means that, on average, the complete vessel call through the port, including time spent in the outer anchorage, takes 74% longer to complete than the baseline trip. The TTI is much lower at Theodore than at River Harbor facilities, suggesting that although trips are on average longer, they are more consistent than trips to River Harbor

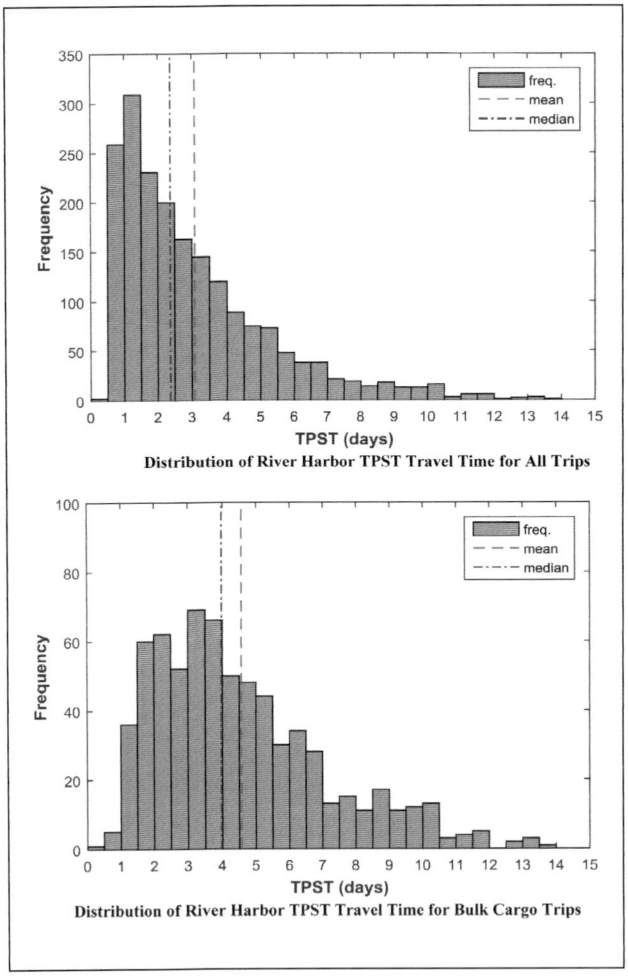

Figure 4. Distribution of River Harbor TPST travel times.

(possibly due to a less diverse set of cargo types and vessel types at Theodore).

PTI values are very large in this analysis due to the amount of variability in trip times through the system based on differences in cargo types, ship size, dock operations, and so forth. The PTI for TPST is 6.19, which means that a vessel has to allow over 500% extra time above the baseline travel time to ensure it will exit the port on time 95% of the time. Some of the variability goes away with the PCT as each vessel has to allow just over 400% additional time due to variability.

It should be noted that the TPST and PCT metrics, which have been derived from approaches used in highway traffic congestion studies, are not perfectly transferable to marine vessel travel times through ports. For example, whereas a vehicle stuck in traffic does not have many options for circumventing the congested areas and reducing overall travel time, the observed in-port times for vessels are quite likely due to decisions being made in real time by the vessel owner, operator, or both vessel owner and operator. Questions remain and future research is needed to pinpoint the degree to which port call durations and associated variability are driven by vessel operator preferences versus congestion (either dockside or due to traffic in the maintained channels), administrative delays (access to pilots, customs officials, etc.), weather issues, or other factors that the vessel operators cannot control.

In roadway mobility analyses, delay is represented as the extra travel time needed beyond the travel time that would happen at freeflow or the baseline. For the purposes of a port analysis, the additional travel time may not be "travel" time but "dock" time or port operations time. The time above baseline on a "per vessel" basis averages about 36.5 h for all vessels in the Port of Mobile. The value is higher in River Harbor operations at 37.5 h compared with vessels at Theodore with 25.1 h, reflecting greater variability in the duration of dock stays at River Harbor than at Theodore.

Mobility Results by Vessel Type

As mentioned previously, the Lloyd's Register database was used to assign attributes such as vessel type and size to the vessels included in the Port of Mobile analysis. This section shows the performance measures for both River Harbor and Theodore split out by vessel type (Table 1). For the Theodore complex, not all vessel types are present. Keep in mind that for calculating each row of this table, all statistics were calculated/recalculated for all of the appropriate vessels such that the River Harbor and Theodore "total" rows have the complete travel time distributions associated with all vessel types, which is why the resulting performance measures may seem out of line with the data for the individual vessel types.

There are no container or crude oil vessels at Theodore. The bulk, general, and crude oil vessels all have longer baseline travel times than the River Harbor complex as a whole. At Theodore, the bulk, chemical, and other vessels have longer baseline travel times than the average for the Theodore complex. These results shed light on the effect of trip distribution tails on the performance measures. The very different results by vessel type (e.g., baseline TPST travel time of 19.3 h for container vessels at River Harbor versus 64.4 h for bulk vessels) demonstrate that the analysis results for the entire port facility, as a whole, may not be nearly as meaningful as the results for analyses into the data using vessel type or vessel size. More analysis will need to fully develop this idea.

At the River Harbor complex, it appears that the container vessels have the shortest TPST and PCT baseline travel times by far over other types of vessels (Table 1). PTI and additional time above baseline per vessel are also lowest for the container vessels. At the Theodore complex,

Table 1. Overall Results for All Vessels by Vessel Type

Trip type and facility	Baseline travel time: 25th percentile (hours)	Travel time index	Planning time index: 95th percentile	Time above baseline travel time (hours)
TPST				
Overall	33.2	1.74	6.19	36.5
River Harbor	32.6	1.80	6.31	37.5
Bulk	64.4	1.56	3.73	55.9
Chemical	31.8	1.15	2.54	11.1
Container	19.3	1.25	2.40	8.0
General	46.5	1.39	3.63	36.0
Crude oil	47.4	1.25	3.96	29.4
Other	31.1	1.71	5.46	39.0
Theodore	35.6	1.38	3.86	25.1
Bulk	56.1	1.37	4.02	45.7
Chemical	39.4	1.27	2.75	23.8
General	23.6	1.21	4.19	8.4
Other	39.0	2.09	2.59	39.4
PCT				
Overall	28.0	1.68	5.09	25.8
River Harbor	27.7	1.73	5.26	26.3
Bulk	48.6	1.61	3.65	37.9
Chemical	28.7	1.15	2.82	9.8
Container	17.4	1.19	1.79	5.2
General	40.8	1.47	3.26	33.0
Crude oil	42.9	1.07	1.51	6.0
Other	28.1	1.57	4.81	30.0
Theodore	92.7	1.46	3.35	19.2
Bulk	54.7	1.39	3.22	43.5
Chemical	33.2	1.28	2.15	15.2
General	22.6	1.22	3.57	7.9
Other	37.9	1.73	2.62	36.4
TTIB				
Overall	2.3	1.07	1.33	0.2
River Harbor	2.3	1.11	1.33	0.3
Bulk	2.3	1.11	1.41	0.3
Chemical	2.3	1.07	1.19	0.6
Container	2.3	1.07	1.22	0.2
General	2.3	1.04	1.22	0.2
Crude oil	2.6	1.05	1.17	0.2
Other	2.3	1.11	1.33	0.3
Theodore	1.8	1.05	1.24	0.0
Bulk	1.8	1.05	1.18	0.0
Chemical	1.8	1.05	1.24	0.1
General	1.7	1.10	1.18	0.0
Other	1.8	1.05	1.24	0.0

Note: TPST = Total Port System Time; PCT = Port Cycle Time; TTIB = Travel Time Inbound.

the general cargo vessels have the shortest TPST and PCT baseline travel times. They also have the lowest TTI and additional time above baseline per vessel; however, they do not have the lowest PTI. The lowest PTI at Theodore belongs to the chemical and other vessels.

Other Mobility Analyses

Several other analyses were performed or are planned but were not shown in this paper due to length constraints. They are documented elsewhere for the interested readers (*21*). These other efforts review changes in mobility levels due to:

- Vessel size, showing that some of the more reliable port mobility was linked to some of the larger cargo vessels.
- Poor weather conditions such as fog, wind, and so on, showing that poor weather conditions prolong anchorage time, port dwell time, or both, thus contributing to poorer mobility performance.

Conclusions

The primary objective of this research was to use archival AIS data to develop and demonstrate a freight fluidity methodology for ports that will allow USACE to evaluate federal navigation projects in terms of their contribution to overall port system fluidity. The research shows that existing data such as archival AIS can be used to report mobility measures and performance.

The researchers successfully developed and presented a methodology for monitoring freight fluidity at ports with consideration to data inputs, data quality control procedures, freight fluidity performance measures, and guidance on the determination of data elements for input to the measures using port data. The paper includes a successful proof-of-concept application of the methodology at the Port of Mobile, Alabama.

Specifically, the researchers identified:

- Data issues associated with the vessel characteristics in the AIS messages such as miscoded information and missing data.
- Port freight fluidity mobility measures and calculation parameters.
- Differences between roadway and waterway mobility measurement, including how "delay" on the waterway system does not have the same meaning as on the roadways—specifically, the researchers developed a measure of "time above the baseline travel time" to capture additional processing time of movements in the port.
- That container vessels, in general, have the shortest stays at the Port of Mobile and have some of the best overall performance.
- That bulk vessels, in general, have the longest stays at the Port of Mobile and often spend 10 days to 2 weeks at the port. It was noted that these long stays in port quite likely are due to vessel operator or shipper-driven logistical preferences, or both, as opposed to congestion issues associated with the port itself. This is an open topic for additional research.

Although this work provides the basis of port freight fluidity work, future work should undertake travel time analyses including other factors affecting port travel times (e.g., vessel size, weather impact, pilot safety guidelines). As indicated earlier in this paper, a key element of freight fluidity is the cost of moving freight shipments from end to end in the supply chain; however, this work did not include cost implications. Future work should investigate costs, perhaps by using tonnage affected by non-ideal travel times as a surrogate for costs. Another opportunity for future work is extending the methods here to additional ports in the United States. Other ongoing efforts such as the Bureau of Transportation Statistics Port Performance Freight Statistics Program (*23*) are seeking new quantitative metrics to objectively evaluate port capacity and throughput on the national Marine Transportation System and the broader freight transportation system.

Acknowledgments

The authors thank the U.S. Army Engineer Research and Development Center for sponsoring this work (Broad Agency Announcement contract award # W912HZ-16-C-0033) entitled "Developing and Implementing a Freight Fluidity Performance Management Framework for U.S. Ports" upon which this paper is based.

Author Contributions

The authors confirm contribution to the paper as follows: study conception and design: Jim Kruse, David L. Schrank, Kenneth N. Mitchell, Patricia K. DiJoseph; data collection: Jim Kruse, Dong Hun Kang, Kenneth N. Mitchell, Patricia K. DiJoseph; analysis and interpretation of results: Jim Kruse, Dong Hun Kang, David L. Schrank, William L. Eisele; draft manuscript preparation: Jim Kruse, David L. Schrank, William L. Eisele, Kenneth N. Mitchell, Patricia K. DiJoseph. All authors reviewed the results and approved the final version of the manuscript.

References

1. Fluidity Indicator Website. *Transport Canada*, https://stats.tc.gc.ca/Fluidity/Login.aspx. Accessed February 8, 2017.
2. Eisele, W. L., L. P. Tardif, J. C. Villa, D. L. Schrank, and T. J. Lomax. Evaluating Global Freight Corridor Performance for Canada. *Journal of Transportation of the Institute of Transportation Engineers*, Vol. 1, 2011, pp. 39–57.
3. *Transportation Research Circular Number E-C187: Developing Freight Fluidity Performance Measures: Supply Chain Perspective on Freight System Performance*. Transportation Research Board of the National Academies, Washington, D.C., 2014.
4. Eisele, W. L., and J. Villa. *Transportation Research Circular E-C207: Freight Fluidity Scale of Analysis*. Transportation Research Board of the National Academies, Washington, D.C., 2015.
5. US Army Corps of Engineers. *Civil Works Direct Program Development Policy Guidance*. Fiscal Year 2019, http://www.publications.usace.army.mil/Portals/76/Publications/EngineerCirculars/EC_11-2-214.pdf?ver=2017-07-26-102815-160.
6. US Army Corps of Engineers. *Navigation Mission*, http://www.usace.army.mil/Missions/Civil-Works/Navigation/
7. US Army Corps of Engineers. *Navigation Program, Services and Projects*, http://www.usace.army.mil/Portals/2/docs/civilworks/5yr_devplan/fy11_5yrplan.pdf
8. International Telecommunication Union. *Technical Characteristics for an Automatic Identification System using Time-division Multiple Access in the VHF Maritime*

Mobile Band. ITU-R M.1371.4. Geneva, Switzerland, 2009.

9. Scully, B., and K. N. Mitchell. *Archival Automatic Identification System (AIS) Data for Navigational Project Performance Evaluation.* ERDC/CHL CHETN-IX-40. US Army Corps of Engineers. August, 2015.
10. Scully, B., and K. N. Mitchell. Underkeel Clearance Reliability Model for Dredged Navigation Channels. *Transportation Research Record: Journal of the Transportation Research Board*, 2017. 2611: 41–49.
11. Farhadi, N., S. Parr, K. N. Mitchell, and B. Wolshon. Use of Nationwide Automatic Identification System Data to Quantify Resiliency of Marine Transportation Systems. *Transportation Research Record: Journal of the Transportation Research Board*, 2016. 2549: 9–18.
12. Mitchell, K. N., and B. Scully. Waterway Performance Monitoring via Automatic Identification System (AIS) Data. *Transportation Research Record: Journal of the Transportation Research Board*, 2014. 2426: 20–26.
13. Scully, B. Management of Coastal Navigation Channels Based on Vessel Underkeel Clearance in Transit. PhD dissertation. Mississippi State University, 2016
14. U.S. Coast Guard Navigation Center. *Nationwide Automatic Identification System*, 2014. https://www.navcen.uscg.gov/?pageName=NAISmain. Accessed July 30, 2014.
15. US Army Corps of Engineers. *Navigation Enews.* Vol. 20, 2011, http://chl.erdc.usace.army.mil/media/1/2/4/6/Nav_enews_Oct_2011.pdf
16. Calder, B., and K. Schwehr. Traffic Analysis for Calibration of Risk Assessment Methods. In *U.S. Hydrographic Conference,* 11–14 May 2009, International Hydrographic Organization, Norfolk, VA, 2009.
17. *Waterborne Commerce of the United States, Calendar Year 2014, Part 2 – Waterways and Harbors Gulf Coast, Mississippi River System and Antilles*, Army Corps of Engineers Institute for Water Resources, 2015, http://www.navigationdatacenter.us/wcsc/webpub14/webpubpart-2.htm. Accessed February 8, 2017.
18. NAFTA Region Port Container Traffic 2014. *American Association of Port Authorities*, 2014, http://www.aapa-ports.org/unifying/content.aspx?ItemNumber=21048. Accessed February 8, 2017.
19. U.S. Port Ranking by Cargo Volume 2014. *American Association of Port Authorities*, http://www.aapa-ports.org/unifying/content.aspx?ItemNumber=21048. Accessed February 8, 2017.
20. U.S. Maritime Administration. *2014 Vessel Calls in U.S. Ports, Selected Terminals and Lightering Areas*, 2014. https://www.marad.dot.gov/resources/data-statistics/. Accessed February 8, 2017.
21. Kruse, J. *DRAFT Technical Memorandum – Port of Mobile System Performance Analysis* (unpublished). Texas A&M Transportation Institute, College Station, TX, November, 2017.
22. US Department of Transportation Federal Highway Administration. *Travel Time Reliability: Making It There on Time, All the Time*, 2017. https://ops.fhwa.dot.gov/publications/tt_reliability/TTR_Report.htm. Accessed February 8, 2017.
23. Bureau of Transportation Statistics. *Port Performance Freight Statistics Program.* https://www.rita.dot.gov/bts/port_performance/freight_statistics_program. Accessed February 8, 2017.

The Standing Committee on Ports and Channels (AW010) peer-reviewed this paper (18-02268).

Article

TRR
JOURNAL OF THE TRANSPORTATION RESEARCH BOARD

Transportation Research Record
2018, Vol. 2672(11) 41–52

DOI: 10.1177/0361198118794066
journals.sagepub.com/home/trr
SAGE

Leveraging the Value of Land and Landside Access to Fund Port Infrastructure in Texas

Rafael M. Aldrete[1], Carl James Kruse[2], David Salgado[1], Sharada R. Vadali[3], Abhisek Mudgal[1], Juan Carlos Villa[3], Lorenzo E. Cornejo[1], and Deog Sang Bae[1]

Abstract

This paper introduces port authority transportation reinvestment zones (TRZs), a funding tool created by the Texas Legislature to help fund long-term economic development port projects. TRZs were first introduced in Texas in 2007 as a tool to allow county and municipal governments to raise funds to help pay for transportation improvements using the property tax mechanism. Since then, the legislation has been amended to include other transportation modes, such as transit, rail, and parking facilities. Amendments passed in 2013 introduced the concept of port authority TRZ, which expanded authorized use of the tool to the state's port authorities and navigation districts. Most of the existing funding tools available to Texas ports focus on the development of infrastructure within port property. Port authority TRZs complement existing funding sources by providing a funding tool flexible enough to fund port projects both inside and outside port property. Port authorities can take advantage of the TRZ concept to join forces with neighboring local governments and the Texas Department of Transportation to fund landside transportation enhancements that improve port accessibility and the regional economy. Currently, there are four active port authority TRZs in the following locations: Port of Beaumont, Port of Port Arthur, Port of Brownsville, and Sabine-Neches Navigation District. This paper summarizes the legal framework of port authority TRZs, describes the role that these TRZs play in port funding and finance, and presents highlights of the port authority TRZs that have already been established throughout the state.

In 2007, the Texas State Legislature created an innovative transportation funding mechanism called the transportation reinvestment zone (TRZ), which allows local governments to set aside local match contributions for the transportation projects most critical to their communities. TRZs generate transportation project funding by capturing and leveraging the real estate/land development value resulting from a transportation project. Development and expansion of transportation projects like highways and transit often spur land development in areas around the project (*1*).

TRZs allow cities or counties to designate an area around a project as an impact zone to capture some or all of the increment in local property and sales tax revenues resulting from the growth in the zone's tax base. That incremental tax revenue is used to support funding and financing of the project by combining it with traditional and nontraditional transportation funding sources (i.e., federal or state funds, tolls, etc.) to assemble a complete project funding package. In this manner, the land development attributable to the project is used to help fund the project. Since TRZ legislation was first enacted to allow the creation of municipal and county TRZs, several municipal and county TRZs have been established in different communities throughout Texas (*1*, *2*).

Municipal and county TRZs are an increasingly important funding source for the expansion and improvement of economically critical links in the U.S. interstate and Texas state highway systems. However, there are comparatively few loan or grant programs that are dedicated specifically to the marine mode, which leads to the need for new and creative approaches to funding port-related infrastructure. There is no source of direct

[1]Texas A&M Transportation Institute, El Paso, TX
[2]Texas A&M Transportation Institute, Houston, TX
[3]Texas A&M Transportation Institute, College Station, TX

Corresponding Author:
Address correspondence to Rafael M. Aldrete: r-aldrete@tamu.edu

state investment in deep or shallow draft ports in Texas (*3*).

In 2013, the Texas Legislature expanded the TRZ legislation to consider port authority TRZs as another method to fund infrastructure in these multimodal facilities. Although the processes for establishing a port authority TRZ are fundamentally the same as conventional TRZs, there are differences in the ways these funds may be utilized. TRZs should be further recognized as multimodal funding tools that enable the revitalization of infrastructure and new development to accommodate the increasing demand on U.S. interstate and Texas state highway systems, rail, and ports.

The objective of this paper is to introduce port authority TRZs, an innovative port infrastructure funding tool. This paper summarizes findings of a component of a research project sponsored by the Texas Department of Transportation (TxDOT) entitled *Tools for Port Authority Transportation Reinvestment Zones (TRZ) and TRZs for Multimodal Applications* (*4*). The paper consists of five major sections. The present section serves as the introduction. The second section describes the legal framework that governs the formation, administration, and operation of port authority TRZs. The third section summarizes funding and financing mechanisms used by port authorities in Texas, the role that TRZ funding plays in this context, and the practical limitations to their use. The fourth section presents highlights of the four port authority TRZs that are currently active in Texas. The last section provides some concluding remarks.

Port Authority TRZ Legal Framework

This section describes the legal framework that governs the formation, administration, and operation of port authority TRZs. First, this section discusses the origin and the evolution of the legislation that governs TRZs from the inception of the law through its most recent changes and the creation of the port authority TRZ as an infrastructure funding tool for ports. Second, the section describes in detail the statutory framework regulating the implementation of port authority TRZs, including their formation and administration, flow of revenues, and interaction with other local government reinvestment mechanisms.

Legislative Background

The creation of the TRZ mechanism dates to Texas legislation first enacted by Senate Bill (SB) 1266 in the 80th Legislative Session in 2007 (*5*). A TRZ is an innovative tool that allows local governments to raise funds to help pay for transportation improvements using all or part of the incremental growth in property and sales taxes from a designated area around the project. SB 1266 amended Chapter 222 of the Transportation Code, and the TRZ provisions appear in Sections 222.105–107. Since then, the legislation for TRZs has been evolving in subsequent legislative sessions as a response to first implementers and their experiences with technical issues that were present in the legislation as originally conceived in SB 1266. SB 1266 laid the foundation for the development of two types of TRZs—municipal TRZs and county TRZs (*1*).

Since 2007, the legislature has considered and approved several bills that modified or expanded the use and types of TRZs, or simply clarified the process or requirements to establish one. During the 83rd Legislative Session in 2013, three bills were filed and enacted into law that extended the TRZ concept to other types of projects or jurisdictions:

- Senate Bill 1747—County Energy TRZs.
- House Bill (HB) 2300—County Energy TRZs.
- Senate Bill 971—Port Authority TRZs.

The first two of these bills, SB 1747 and HB 2300, dealt with the creation of a new type of county TRZ called county energy transportation reinvestment zones (CETRZs). CETRZs are aimed at helping counties in shale energy regions alleviate pavement deterioration to roads, bridges, and other infrastructure caused by oil and gas exploration (*6*, *7*). In 2017, SB 1305 repealed the CETRZs in the 85th Legislative Session (*8*).

In contrast, SB 971 expanded the TRZ concept to all modes of transportation by enabling port authorities and navigation districts to establish port authority TRZs (*9*). Although SB 971 amended only the Texas Transportation Code, there are several already existing statutes that affect the planning, implementation, and financing of port authority TRZs, such as the state's water, local government, and tax codes. These different laws, statutes, and regulations provide the legal framework of port authority TRZs. The sections that follow describe the changes to the Texas Transportation Code introduced by SB 971 to enable the creation of port authority TRZs and discuss other state laws and statutes that govern or influence their planning or implementation. The terms "port authority" and "navigation district" are used interchangeably and describe the same entity.

Port Authority TRZ Definitions

Port authorities in Texas can assume several legal forms. However, all of the port authorities presented in this paper are navigation districts with taxing authority that handle cargo shipments at their facilities or facilitate cargo handling activities. Because TRZs rely on property tax increment revenues, only port authorities with taxing authority are able to use the mechanism. SB 971 Section

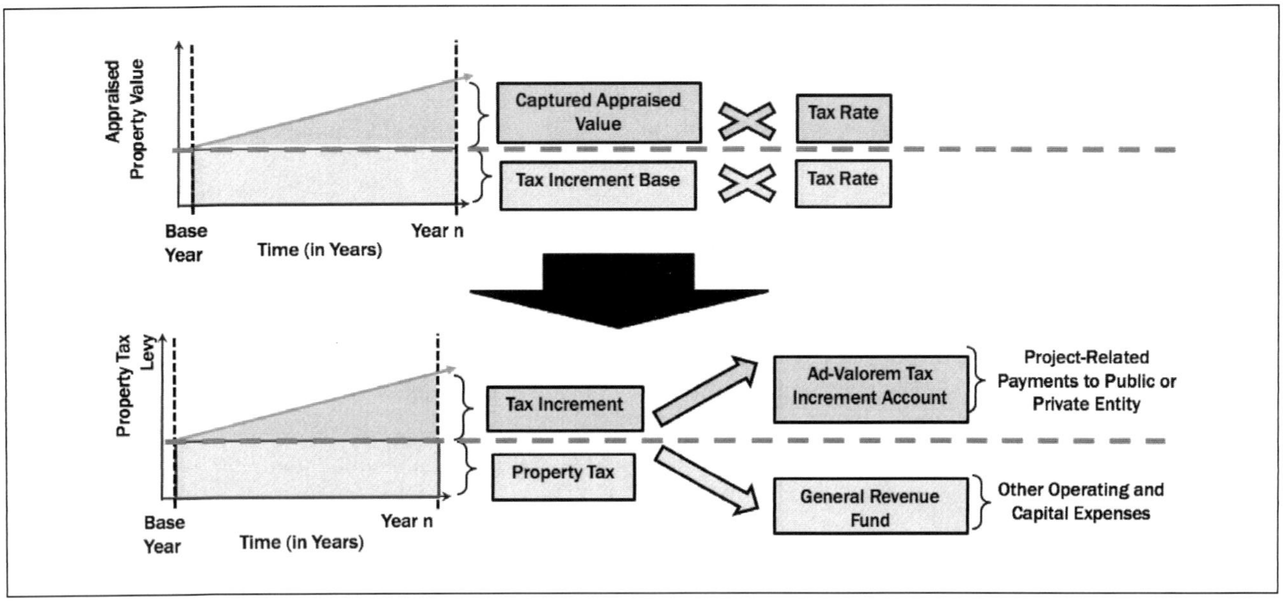

Figure 1. Port authority TRZ definitions (*10*).

222.1075 provides basic TRZ definitions applicable to port authorities for concepts such as (*a*) tax increment, (*b*) captured appraised value, and (*c*) tax increment base. These port authority TRZ concepts are very similar to those applicable to a traditional county TRZ (*9*). The definitions are included below and illustrated in Figure 1.

- **Tax Increment Base.** The tax increment base of a port authority is defined as the total appraised value of all real property taxable by the port authority and located in a TRZ for the year in which the zone was designated.
- **Captured Appraised Value.** The captured appraised value of real property taxable by a port authority for a year is defined as the total appraised value of all real property taxable by the port authority and located in a TRZ for that year less the tax increment base of the port authority.
- **Annual Tax Increment.** The amount of a port authority's tax increment for any given year is defined as the amount of *ad valorem* taxes levied and collected by the port authority (or by the commissioners' court on behalf of the port authority) for that year on the captured appraised value of real property taxable by the port authority and located in a TRZ.

Formation and Administration of Port Authority TRZ

As noted by Cassidy et al., the formation and administration of a port authority TRZ under SB 971 is very similar to the existing TRZ process available for cities and counties (*11*). This process consists of five steps (see Figure 2). The following paragraphs explain in detail each step.

Step 1. Initiation. The process of implementing a port authority TRZ starts with the initiation stage. This stage consists of selecting an eligible target area for TRZ funding that is unproductive or underdeveloped as required by SB 971. The funds generated by the TRZ should be used for a project that is necessary or convenient for the proper operation of a maritime port or waterway and that will improve security, movement, and intermodal transportation of cargo or passengers in commerce and trade, as described by Texas Transportation Code Section 222.1075 (a)-(3). A preliminary increment capture analysis to assess port authority TRZ revenue needs to be performed in this step. The results of this preliminary assessment could be used to initiate a dialog with stakeholders, such as navigation district board commissioners, county commissioners, TxDOT, metropolitan planning organizations, and other stakeholders, to build support. The identification of a TRZ champion facilitates this process because it requires a significant amount of interagency collaboration.

Step 2. Zone Formation. This second stage of the TRZ implementation process involves three substeps: (*a*) determination of zone boundaries, (*b*) provision of a 60-day notice, and (*c*) refinement of the TRZ increment capture analysis conducted in Stage 1.

- **Determination of Zone Boundaries**. The first step is to define zone boundaries and identify the

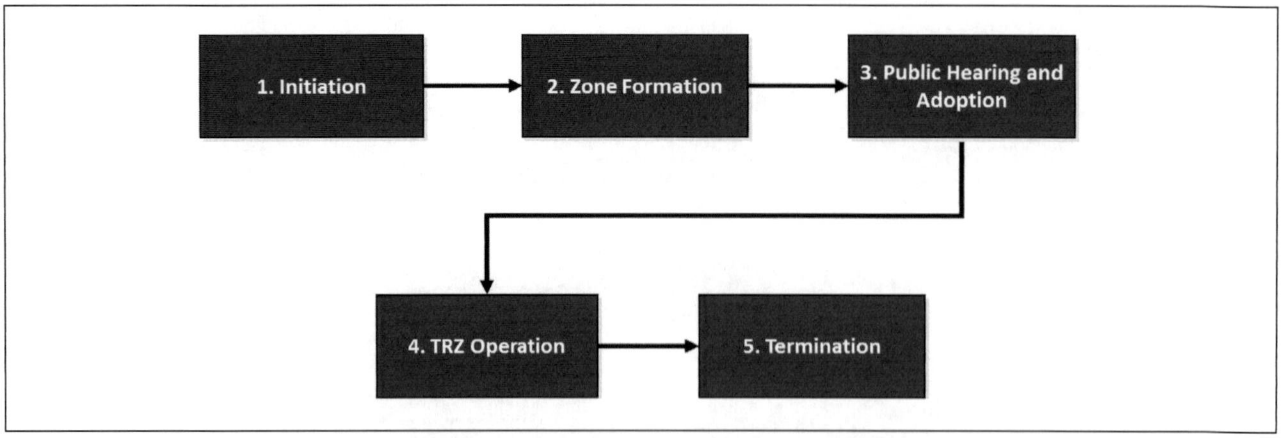

Figure 2. Implementation stages for port authority TRZs.

parcels that will be within the port authority TRZ. The port authority determines zone boundaries by designating a contiguous geographic area within its jurisdiction as a TRZ under Section 222.1075 (c). The zone should have its boundaries defined with sufficient definiteness to identify them with ordinary and reasonable certainty (Section 222.1075 [f]-[1]). In addition, the parcels within the TRZ boundaries should be identified and listed. Consistent with the requirements for municipal and county TRZs, SB 971 establishes that any previous property tax relief and abatement agreements authorized by the port authority (including those under Chapter 312 of the Tax Code) are to be excluded (deducted) from the annual tax increment paid into the tax increment account.

- **Provision of 60-Day Notice**. There is a 60-day notice period that needs to take place before the TRZ is designated. The legislation specifies that the port commission hold a public hearing on the creation of the zone no later than the 30th day before the date when the TRZ is expected to be designated.
- **Refine TRZ Increment Capture Analysis**. During this 60-day period, the preliminary increment capture analysis conducted in the initiation stage should be refined to ensure that the revenue will be sufficient to meet the expected TRZ-related financial commitments (e.g., debt service) entered into by the navigation district. This refining process includes generating the highest possible cadastral (parcel) data and refining assumptions related to pace of development and property values. The refined increment capture analysis should estimate the potential tax increment revenue potential within the proposed TRZ boundaries.

Step 3. Public Hearing and Adoption. After setting up initiation and zone formation, a public hearing notice should take place before the TRZ is designated. The public hearing should be held no later than the 30th day from the date the intent to create the zone is announced. A notice must be published in a newspaper no later than the seventh day preceding the hearing day, as described in Section 222.1075 (e). Once the TRZ is designated, the port authority establishes or designates the TRZ by order or resolution.

Step 4. TRZ Operation. Every subsequent year after the port authority TRZ base year, the annual TRZ tax increment will be transferred into the navigation district's TRZ tax increment account, as described in Section 222.1075 (h)-(1) and illustrated in Figure 3. The annual tax increment may be used for repaying any loan or other debt incurred to finance a port project in the zone, as defined in Section 222.1075 (h)-(2).

Once TRZs are established, monitoring and evaluation are critical steps. Although not required by the law, port authorities will find it in their interest to establish monitoring and evaluation of TRZ revenues to optimize revenue and payment streams. This information may be used for multiple purposes. For example, when revenue from new development may be lower than expected, the established monitoring and evaluation processes enable a port administration to develop insights in order to activate a contingency plan.

Step 5. Termination. According to SB 971, a port authority TRZ terminates on December 31 of the year in which the port authority completes any contractual requirement that included the pledge of TRZ revenue. In addition, a port authority TRZ terminates on December 31 of the 10th year after the year the zone was designated if the

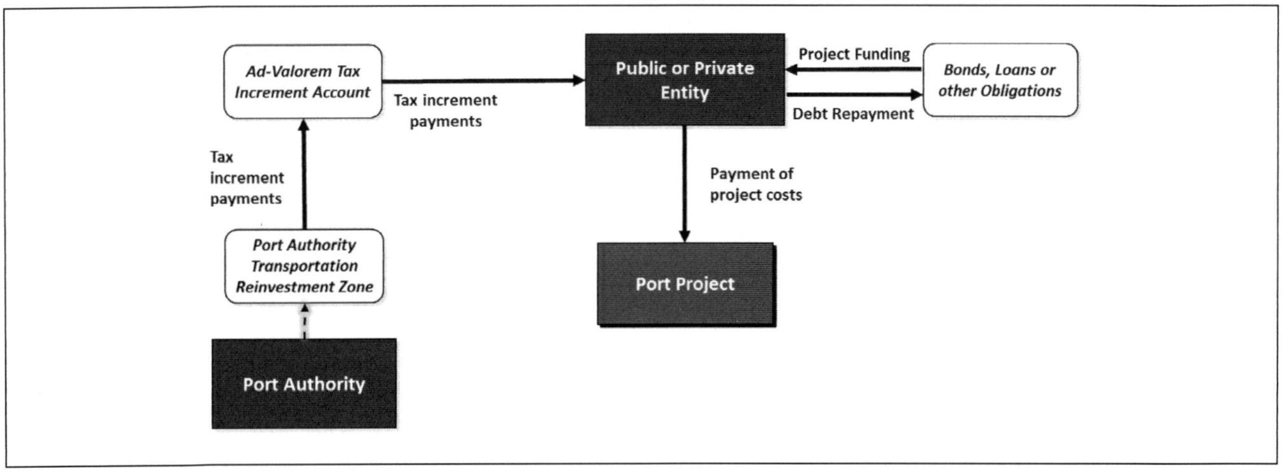

Figure 3. Flow of funds in port authority TRZ financing.

port authority has not used the zone for the purpose for which it was designated before that date.

Funding and Financing Mechanisms Used by Port Authorities in Texas

This section describes the funding mechanisms and financing methods currently used by port authorities in Texas. First, this section discusses funding sources available at the state, federal, and local level for port authorities to pay for their investment and operational needs. Second, this section describes the financing methods that Texas port authorities typically use to leverage their funding sources and secure the capital needed for infrastructure investment.

Funding Mechanisms

The five main local, state, and federal funding mechanisms available to Texas ports are as follows: (*a*) direct and indirect funding tools provided by the state legislature; (*b*) Transportation Investment Generating Economic Recovery (TIGER) grants; (*c*) Fostering Advancements in Shipping and Transportation for the Long-Term Achievement of National Efficiencies (FASTLANE) program grants; (*d*) port security grants; and (*e*) taxable property values. The paragraphs that follow describe each of these mechanisms.

Direct and Indirect Funding Tools Made Available by Texas Legislature. The 77th Texas Legislature created one direct funding vehicle—the Port Access Account Fund—in 2001 to provide funds for Texas ports to finance security improvements, infrastructure projects, and related studies. However, the fund did not survive the bill, which abolished and consolidated certain funds for the 77th Legislature and never actually allocated funds (*3*).

TIGER Grants. TIGER grants are awarded through a competitive selection process that has allocated $5.1 billion nationally since 2009 for projects that "promise to achieve critical national objectives" and "have a significant impact on the Nation, a region, or a metropolitan area." Awarded projects are also expected to have a state or local match. The program is currently in its eighth round (*12*), which has been funded at $500 million.

FASTLANE Program Grants. The Fixing America's Surface Transportation Act (FAST Act) established the Nationally Significant Freight and Highway Projects (NSFHP) program to provide federal financial assistance to projects of national or regional significance and authorized the program at $4.5 billion for fiscal year (FY) 2016 through 2020, including $800 million for FY 2016 to be awarded by the Secretary of Transportation. The grants from the NSFHP program are referred to as FASTLANE grants. The program provides dedicated, discretionary funding for projects that address critical freight issues facing our nation's highways and bridges and for the first time in the U.S. Department of Transportation's (U.S. DOT's) 50-year history, establishes broad, multiyear eligibilities for freight infrastructure. The grants are awarded by U.S. DOT on a competitive basis to projects of national or regional significance that meet statutory requirements (ports are included). NSFHP grants may be used for construction, reconstruction, rehabilitation, property acquisition (including land related to the project and improvements to the land), environmental mitigation, construction contingencies, equipment acquisition, and operational improvements directly related to system performance. In addition to the FASTLANE grants, the act allows states to obligate up to 10% of their total freight apportionment for intermodal or freight rail projects, which specifically includes ports. To date, no funds have been awarded to a Texas entity (*13*).

Port Security Grants. The purpose and permitted uses of port security grant funds have evolved over time. In FY 2005, Congress appropriated funds to specifically address physical security enhancements for critical national seaports. The allowable uses of the funds have been modified on multiple occasions since then, and the funds are no longer used solely for the purchase of physical assets. According to the latest (FY 2017) guidance published by the Federal Emergency Management Administration, port security grant funds are intended to improve port-wide maritime security risk management; enhance maritime domain awareness; support maritime security training and exercises; and maintain or reestablish maritime security mitigation protocols that support port recovery and resiliency capabilities (*14*).

Taxable Property Values. The ability to generate tax revenue or raise funds is directly proportional to the total appraised value of property within the port's taxing jurisdiction (i.e., its tax base). Within the Texas port system, there is a vast range of total appraised values within each port's taxing jurisdiction. These disparities in total appraised value make it much more difficult for ports with smaller tax bases to raise funds than for ports with larger tax bases.

Financing Mechanisms

Aside from cash flow financing (including both fees and taxes), the four primary financing mechanisms used by Texas ports are general obligation bonds, revenue bonds, notes payable, and capital leases.

General Obligation Bonds. A general obligation bond is a bond that is secured by the taxing and borrowing power of the entity issuing it. Voters must approve such bonds (and the authority to levy ad valorem property taxes to cover debt service) before such bonds may be issued. For the system as a whole, there has been a marked decrease in the overall outstanding balance of general obligation bonds. The use of general obligation bonds is typically driven by local conditions, primarily (*a*) the expectation (or lack thereof) of improved business activity that will generate increased economic activity at the port, and (*b*) the willingness of the local population to tax itself.

Revenue Bonds. Revenue bonds are debt securities that have a defined source of anticipated funds to pay both the principal and the interest. These funds may come from an activity, project, or revenue source that is not related to the capacity to levy taxes. From an investor's perspective, these bonds are riskier than general obligation bonds. In many cases, revenue bonds carry a requirement for the ratio of net revenues to revenue bond debt service to be a certain amount or greater. Typically, this ratio is in the range of 1.25 to 1.5. In simple terms, net revenue is net income without depreciation and without any general obligation tax receipts. Not all the ports have revenue bonds outstanding, and of those that do, not all report their ratios. In 2014, three ports reported their ratio: Freeport—2.25:1 (*15*), Beaumont—4.92:1, and Brownsville—28.09:1 (*16*). These ratios indicate that the ports are generating more than enough cash flow from operations to satisfy bond indenture requirements.

Notes Payable. Only one port in Texas, Freeport, reported notes payable as a financing tool in the 2014 financials (Brownsville paid off a note that was reported in 2004) (*15*). Notes payable are simply a loan from a bank or other lending institution that is typically collateralized with the asset being financed.

Capital Leases. A capital lease is a lease agreement in which the lessor agrees to transfer the ownership rights to the lessee after the completion of the lease period. A capital lease is considered to have the economic characteristics of asset ownership. A capital lease would be considered as a purchased asset for accounting purposes. Only one port in Texas, the Port of Corpus Christi, had a capital lease on the books as of 2014 (*17*).

Port Authority TRZ Roles

The existing funding tools available to port authorities and navigation districts are all focused on the development of infrastructure on property owned by them. Port authority TRZs can enable ports to develop port infrastructure on property they do not own by providing financial resources above and beyond operating income and regular tax revenues. The need to develop infrastructure on non-port property was the primary motivation for the creation of the three TRZs in Jefferson County—specifically, the need to provide local match funds for the widening and deepening of the Sabine-Neches Waterway.

Port authorities already use a variety of mechanisms to influence development near the port. Mechanisms such as free trade zones, tax abatements, and special purpose districts influence business operating costs (for businesses located at ports) by altering taxes but are not

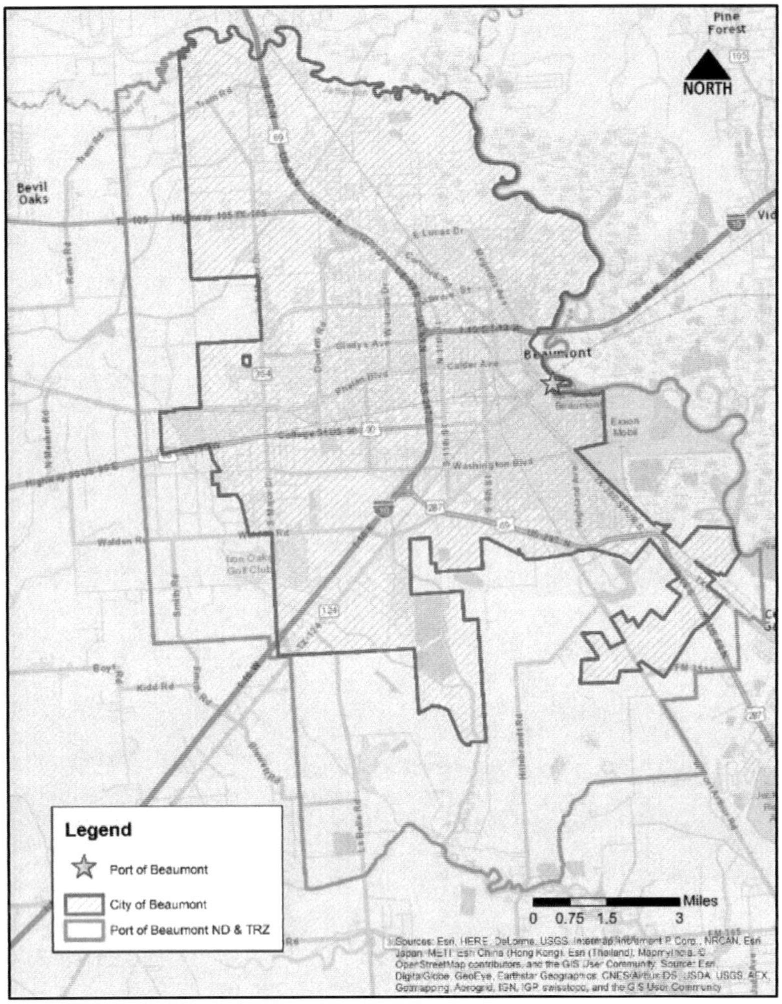

Figure 4. Port of Beaumont and TRZ boundaries.

directly geared toward the development of port infrastructure specifically. Indirectly, however, the growth of these businesses adds value and revenues to ports, which have traditionally been used to finance port investments via bonding. What is missing is a mechanism for port authorities to develop infrastructure in areas adjacent to the port that will have a direct influence on the port's operating capacity and efficiency, such as improved access to and mobility around the port.

The port authority TRZ mechanism could significantly help port authorities address infrastructure needs in areas adjacent to port property that can improve the port's efficiency or security. This tool could allow navigation districts to address last-mile connectivity issues and capacity constraints without placing an undue burden on the districts' financial condition.

Port authority TRZs make it possible for port authorities to team with counties or cities that also have TRZs to construct projects that will benefit all parties. When it is politically and economically expedient to do so, multiple entities can all dedicate their TRZ resources to one project or set of projects that will enhance the transportation infrastructure within the overlapping TRZ designated area and promote economic development.

Practical Limitation of the Port Authority TRZ

There is one practical limitation to port authority TRZs when it comes to targeting infrastructure that is specifically port related. Several TRZs around the state have used their tax increment funds to acquire funding from the State Infrastructure Bank (SIB). However, because

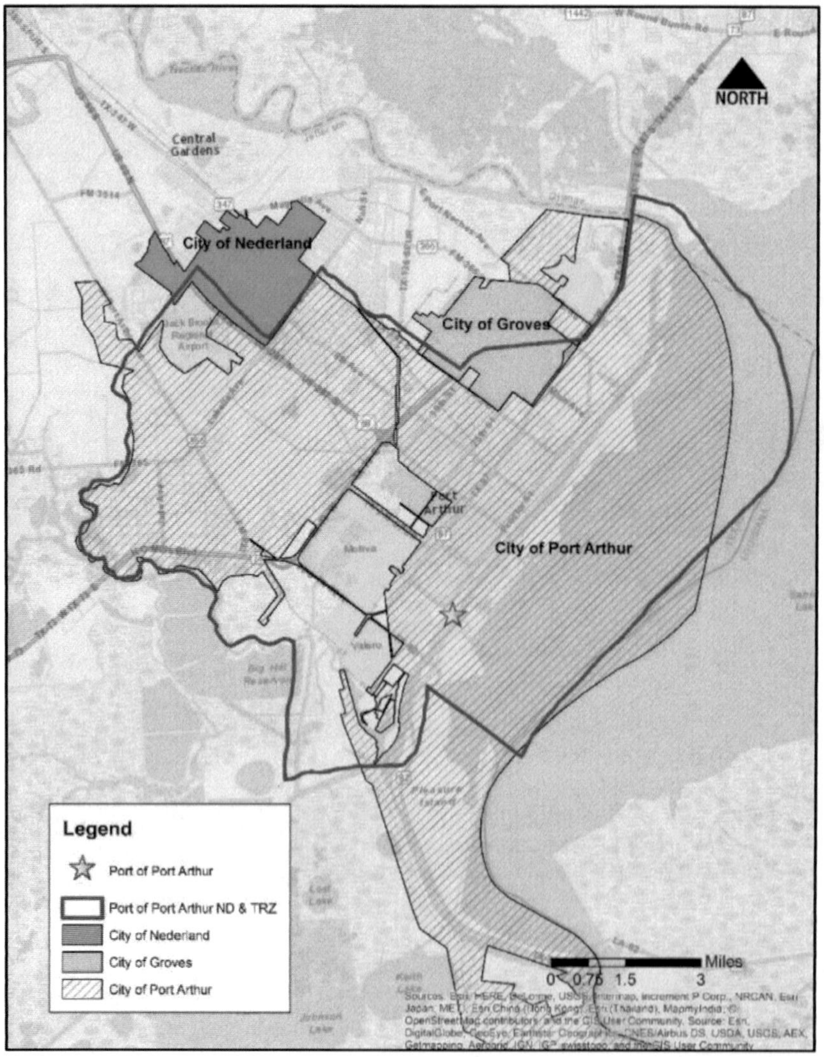

Figure 5. Port of Port Arthur and TRZ boundaries.

the bank's funding comes from the federal and state government's highway programs, the SIB cannot currently be used to finance port projects because port property is not considered part of the state highway system. In such cases, it would be technically feasible to go to the bond market or other lenders, but to date the cost of such financing has been prohibitive.

Active Port Authority TRZs in Texas

Currently, there are four active port authority TRZs in the state of Texas. These port authority TRZs are located in Port of Beaumont, Port of Port Arthur, Sabine-Neches Navigation District (SNND), and Port of Brownsville. The first three port authority TRZs are all situated within Jefferson County. The fourth port authority TRZ is located in Cameron County, which is the southernmost county in the state and borders Tamaulipas, Mexico. The following paragraphs provide specific details for the geographical location of each port authority TRZ and the purpose of establishing them.

Port of Beaumont TRZ

The Port of Beaumont is a deep-water international public seaport located in Jefferson County, Texas, on the Neches River 40 miles inland from the Gulf of Mexico. The port is accessible from the Gulf of Mexico and the Gulf Intracoastal Waterway by means of the federally maintained Sabine-Neches Waterway. The Port of

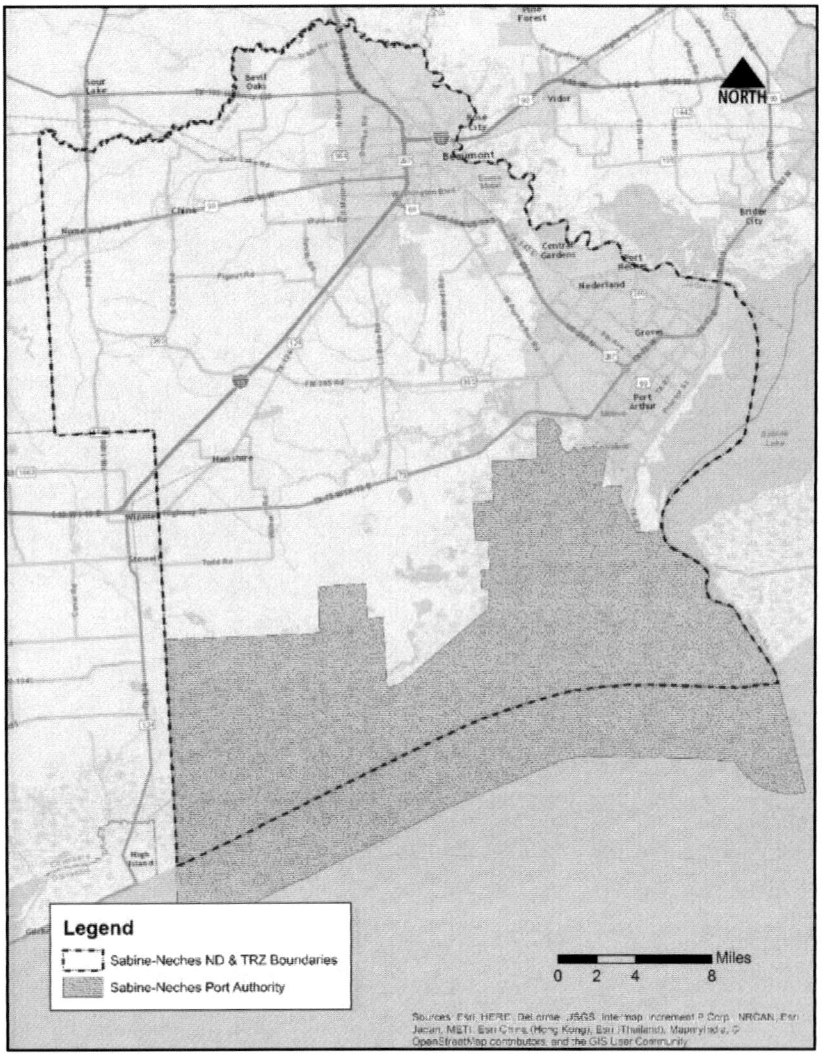

Figure 6. SNND and TRZ boundaries.

Beaumont Navigation District covers an area of about 150 mi^2 including the City of Beaumont, Texas (*18*).

The Port of Beaumont established the Port of Beaumont TRZ No. 1 in December 2013. To date, the Port of Beaumont has not defined the duration of the TRZ. Although the Port of Beaumont TRZ was established to fund improvements in the Sabine-Neches Waterway, the specific improvements have not been defined yet. The Port of Beaumont TRZ has the same boundaries as the navigation district (see Figure 4).

Port of Port Arthur TRZ

The Port of Port Arthur is strategically located midway on the barge shipping routes of the intercoastal waterway that extends from St. Marks, Florida, to Brownsville, Texas. The Sabine-Neches Ship Channel is part of that system, with channel depths to safely accommodate large petrochemical tankers and cargo ships. The port is 19 mi from the Gulf of Mexico and has 24-h channel and dock accessibility (*19*).

Port of Port Arthur Commissioners established the Port of Port Arthur TRZ No. 1 in December 2013. To date, the port has not defined the duration of the TRZ. The port already identified some of the potential uses of the TRZ revenue, including but not limited to Berth 6 expansion, Berths 1–5 modernization, and other potential multimodal transportation improvements. Additionally, TRZ revenue may be used to provide the required local match for deepening the Sabine-Neches

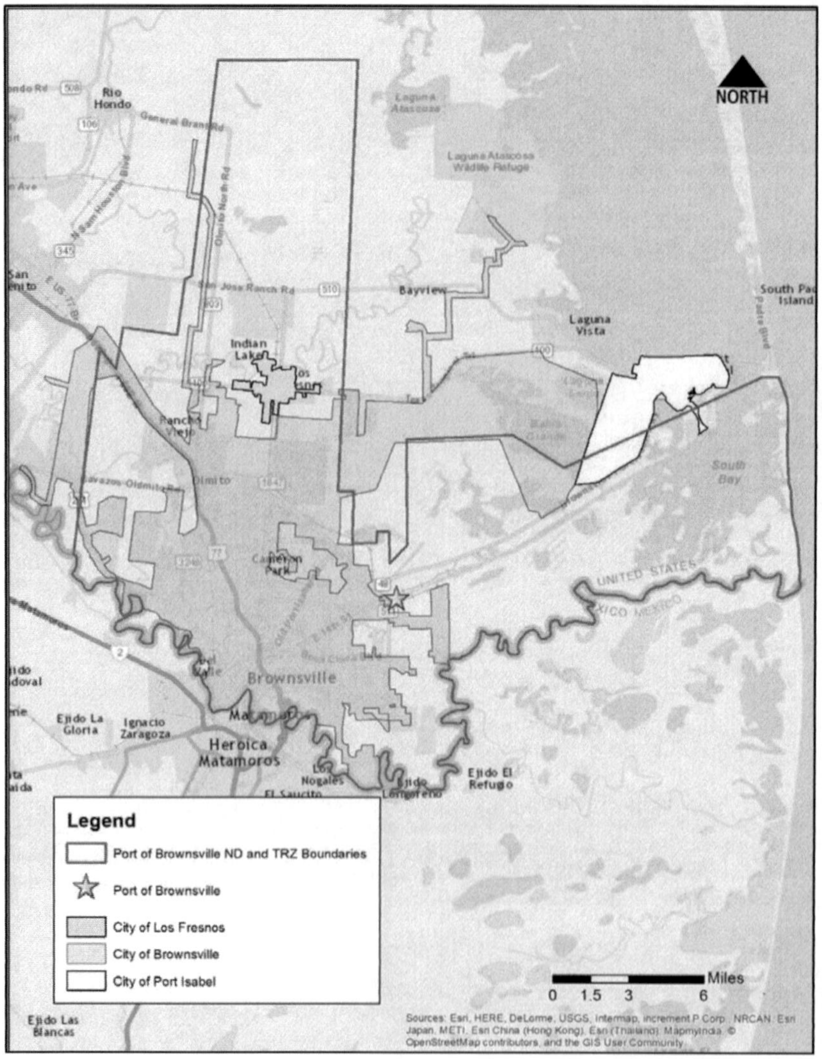

Figure 7. Port of Brownsville and TRZ boundaries.

Navigation Waterway (SNNW) from 40 ft to 48 ft, which is estimated to cost $1.2 billion, of which approximately $366 million must be paid from nonfederal funds. The Port of Port Arthur TRZ has the same boundaries as the navigation district and is shown in Figure 5.

Sabine-Neches Navigation District TRZ

SNND is not a port facility but a unique organization among the state of Texas' navigation districts. It was created to serve as the nonfederal sponsor for dredging activities on the Sabine-Neches Waterway, which serves the Ports of Orange, Beaumont, and Port Arthur.

SNND established the Sabine-Neches Navigation District TRZ No. 1 in December 2013. To date, the navigation district has not defined the duration of the TRZ. TRZ revenue will be used to provide the required local match for deepening the SNNW from 40 ft to 48 ft, which as noted previously, is estimated to cost $1.2 billion, approximately $366 million of which must be paid by nonfederal sponsors. The SNND TRZ has the same boundaries as Jefferson County, as shown in Figure 6. The Water Resources Reform and Development Act bill and the findings of numerous economic studies show that the waterway is the economic engine of the entire county.

Port of Brownsville TRZ

The port is located at the southernmost tip of Texas at the westernmost terminus of a 17-mi channel that flows into the Gulf of Mexico at the Brazos Santiago Pass. The City of Brownsville is 2 mi to the southwest and lies adjacent to the Rio Grande River, providing a convenient gateway via three international bridges to Mexico and beyond (*20*).

The Port of Brownsville established the Port of Brownsville TRZ No. 1 in December 2013. To date, the Port of Brownsville has not defined either the duration of the TRZ or the transportation project that is intended to be funded. The Port of Brownsville TRZ has the same boundaries as the navigation district (see Figure 7).

Conclusions

Current existing funding mechanisms and financing methods available to Texas ports focus on the development of infrastructure within port property. Port authority TRZs complement existing funding sources by providing a funding mechanism flexible enough to fund port projects both inside and outside port property. Additionally, this mechanism can be combined with traditional and nontraditional transportation funding sources (i.e., federal or state funds, tolls, etc.) to assemble a complete project funding package.

Currently, four port authority TRZs have been established in the following locations within the state of Texas: Port of Beaumont, Port of Port Arthur, Port of Brownsville, and SNND. To date, the Port of Brownsville has not identified any transportation project to be funded. Meanwhile, the Port of Beaumont, Port of Port Arthur, and SNND intend to fund improvements to the Sabine-Neches Waterway.

The order of magnitude of funding that the TRZ mechanism can potentially generate can be illustrated using the example of the Port of Brownsville TRZ No. 1. This TRZ consists of 164,441 parcels totaling over 674,000 acres and representing a tax base of $14.37 billion in 2013. Over a period of 35 years, this TRZ could potentially generate a present value of revenue close to $19.9 million (or a total of $59.75 million in nominal revenues) (*4*).

Acknowledgments

The research team acknowledges the support of TxDOT in funding and conducting the research project that led to this paper.

Author Contributions

The authors confirm contribution to the paper as follows: study conception and design: Rafael M. Aldrete, Carl James Kruse, David Salgado, Sharada R. Vadali; data collection: Rafael M. Aldrete, Carl James Kruse, David Salgado, Lorenzo E. Cornejo, Deog Sang Bae; analysis and interpretation of results: Rafael M. Aldrete, Carl James Kruse, David Salgado, Abhisek Mudgal, Juan Carlos Villa; draft manuscript preparation: Rafael M. Aldrete, Carl James Kruse, David Salgado. All authors reviewed the results and approved the final version of the manuscript.

References

1. Aldrete, R. M., S. Vadali, D. Salgado, L. Cornejo, S. Chandra, and A. Bujanda. *Transportation Reinvestment Zones Final Report*. Texas A&M Transportation Institute Policy Research Center, Austin, Texas, 2016.
2. Vadali, S. R., R. M. Aldrete, and A. Bujanda. Financial Model to Assess Value Capture Potential of a Roadway Project. *Transportation Research Record: Journal of the Transportation Research Board*, 2009. 2115: 1–11.
3. Kruse, C. J. *Survey of State Funding Practices for Coastal Port Infrastructure*. Transportation Policy Research Center, 2015. http://d2dtl5nnlpfr0r.cloudfront.net/tti.tamu.edu/documents/PRC-15-51-F.pdf. Accessed 2015.
4. Aldrete, R. M., S. Vadali, C. J. Kruse, D. Salgado, A. Mudgal, J. C. Villa, L. Cornejo, and D. S. Bae. *Tools for Port Authority Transportation Reinvestment Zones (TRZ) and TRZs for Multimodal Applications*. Publication FHWA/TX-17/0-6890-1. Texas A&M Transportation Institute, College Station, Texas, 2016.
5. Texas Legislature. Senate Bill No. 1266. Texas Legislature Online, September 1, 2007. http://www.legis.state.tx.us/billlookup/text.aspx?LegSess=80R&Bill=SB1266. Accessed July 30, 2017.
6. Texas Legislature. Senate Bill No. 1747. Texas Legislature Online, May 2013. http://www.legis.state.tx.us/tlodocs/83R/billtext/html/SB01747F.htm. Accessed July 28, 2016.
7. Texas Legislature. House Bill No. 2300. Texas Legislature Online, June 2015. http://www.legis.state.tx.us/tlodocs/83R/billtext/html/HB02300I.HTM. Accessed July 28, 2016.
8. Texas Legislature Online. Senate Bill No. 1305. April 4, 2017. http://www.legis.state.tx.us/BillLookup/Text.aspx?LegSess=85R&Bill=SB1305. Accessed July 31, 2017.
9. Texas Legislature. Senate Bill No. 971. Texas Legislature Online, May 2013. http://www.capitol.state.tx.us/tlodocs/83R/billtext/html/SB00971F.htm. Accessed November 2015.
10. Vadali, S., R. M. Aldrete, A. Bujanda, S. Samant, Y. Li, T. Geiselbrecht, B. Kuhn, S. Lyle, M. Zhang, K. Dalton, and S. Tooley. *Transportation Reinvestment Zone Handbook*. Texas A&M Transportation Institute, College Station, Texas, 2012 (Update).
11. Cassidy, C. B., L. F. Winland, and B. O'Reilly. Texas Transportation Legislation: Overview of the 84th Legislative Session. Locke Lord LLP, June 26, 2015. http://www.lockelord.com/newsandevents/publications/2015/07/~/media/9D0B96026C554CA2BF7B7082597F134C.ashx. Accessed July 19, 2016.
12. United States Department of Transportation. About TIGER Grants. December 30, 2016. https://www.transportation.gov/tiger/about. Accessed July 30, 2017.
13. United States Department of Transportation. Fixing America's Surface Transportation Act or "FAST Act." February 8, 2017. https://www.fhwa.dot.gov/fastact/factsheets/fastlanegrantsfs.cfm. Accessed July 30, 2017.

14. Federal Emergency Management Administration. Fiscal Year (FY) 2017 Port Security Grant Program (PSGP). Notice of Funding Opportunity (NOFO), Washington, D.C., 2017. https://www.fema.gov/media-library-data/1496328927518-cfc5486f1846263e697754eb2e7fed50/FY_2017_PSGP_NOFO_508.pdf. Accessed July 30, 2017.
15. Port Freeport. Comprehensive Annual Financial Report for the Fiscal Years Ended, September 30, 2014 and 2013, 2014. http://www.portfreeport.com/annual_files/2014Report.pdf. Accessed October 2015.
16. Brownsville Navigation District of Cameron County, Texas. Comprehensive Annual Financial Report for the Year Ended, December 31, 2014. http://www.portofbrownsville.com/wp-content/uploads/2017/01/BND-CAFR-2014.pdf. Accessed July 30, 2017.
17. Port of Corpus Christi. Comprehensive Annual Financial Report for the Years Ended December 31, 2015 and 2014, 2015. http://www.portofcc.com/images/pccpdfs/annual/2015%20CAFR(Web).pdf. Accessed July 30, 2017.
18. Port of Beaumont. About Us. http://www.portofbeaumont.com/. Accessed July 30, 2017.
19. Port of Port Arthur. About the Port. http://portofportarthur.com/international-cargo-shipping-the-port-of-port-arthur/located-on-the-intercoastal-waterway/. Accessed July 30, 2017.
20. Port of Brownsville. About the Port. http://www.portofbrownsville.com/. Accessed July 30, 2017.

The Standing Committee on Ports and Channels (AW010) peer-reviewed this paper (18-05320).

Article

TRR
JOURNAL OF THE TRANSPORTATION RESEARCH BOARD

Transportation Research Record
2018, Vol. 2672(11) 53–64

DOI: 10.1177/0361198118793328
journals.sagepub.com/home/trr
SAGE

Evaluation of the Health Impacts of Rolling Back a Port Clean Trucks Program

Dana Rowangould[1], Gregory Rowangould[2], and Debbie Niemeier[3]

Abstract

Air pollution emissions from marine port truck activity can pose significant health risks to communities living near ports. One approach to reducing these risks is requiring the use of low-emission trucks at port facilities. This analysis evaluates the predicted air quality impacts of the recent revision (rollback) of a Clean Truck Program at the Port of New York & New Jersey. We use the best publicly available information to estimate emissions, concentrations, and health risks at a fine spatial scale across eight counties in New Jersey in 2017 and 2018, when the impacts of the revision of the Clean Truck Program will likely be greatest. We find that as a result of the revision of the program, port truck $PM_{2.5}$ emissions will be higher by seven to eight fold each year, concentrations will increase by up to 1 $\mu g/m^3$, and the risk of death due to exposure to a year's emissions will increase by over 25 in a million in some areas. These effects are estimated in 2017 and 2018 but are expected to continue, gradually tapering off, for at least 15 years into the future. The magnitude of estimated effects of the revised program would be greater if this study had included a broader geographic region, a longer period of time, and a wider range of truck pollutants, exposure pathways, and health outcomes.

Marine port-related activities result in air pollution emissions including diesel exhaust and particulate matter, which are known to pose health risks (*1*). Port-related emissions can affect adjacent communities and contribute to regional pollution levels (*2*). Transportation emissions associated with port truck activity can raise pollution concerns for communities living near ports and along affected truck routes (*3–6*).

Strategies to reduce the effects of port-related pollution in adjacent communities include off-site mitigations such as air filtration, planting vegetation, building hard barriers, and separating port activities from communities. On-site mitigations may include reducing port-related emissions via reduced activity (e.g. limited idling), the use of cleaner fuels, and technological improvements to vehicles or engines.

One way to achieve technological improvements is to encourage or require retrofits or replacements. When applied to port drayage trucks, these clean vehicle policies have the potential to reduce emissions on-site at port facilities as well as along truck routes outside of port boundaries, which may be located in closer proximity to residents. Mandatory clean truck programs at the California ports of Oakland, Los Angeles, and Long Beach have yielded dramatic reductions in port truck emissions, ranging from 30% to 96% (*7–10*), while voluntary programs elsewhere in the U.S.A. have brought more modest reductions, ranging from 1% to 4% (*11*). An analysis of the health impacts of the Clean Truck Program at the ports of Los Angeles and Long Beach estimated that it saved 62 lives in 2012 alone (*10*).

Although clean truck policies have been adopted in some areas, they often face opposition. For example, in April 2017, California passed a transportation funding law which included a provision that may limit the power of air quality districts and ports to regulate truck technology (*12*), although as of July 2017 the ports of Los Angeles and Long Beach continued to plan for the use of cleaner trucks (*13*). Additionally, in 2016 the Port of New York & New Jersey significantly scaled back its Clean Truck Program in light of concerns about the cost

[1]Sustainable Systems Research, LLC, Davis, CA
[2]Department of Civil Engineering, University of New Mexico, Albuquerque, NM
[3]Civil and Environmental Engineering, University of California, Davis, CA

Corresponding Author:
Address correspondence to Dana Rowangould:
dana@sustainablesystemsresearch.net

of truck replacements and the potential for subsequent truck shortages. The program was opposed by the trucking industry, which would have shouldered the majority of these costs.

Estimating the pollution and health outcomes of clean truck programs can inform the design and implementation of pollution reduction plans for ports. While the effects of California's clean truck policies are well understood, the effects of similar policies (or policy reversals) at large ports in other regions have not been evaluated. Additionally, the lone analysis of the health effects of a California clean truck program evaluates outcomes at the block group level (*10*), so it may fail to capture higher exposure levels that are better characterized by parcel- or block-level analyses (*5*).

This analysis evaluates the predicted effects on air quality and health of the recent revision (rollback) of a Clean Truck Program at the Port of New York & New Jersey. We use the best publicly available information to estimate emissions, concentrations, and health risks at fine spatial scales across eight counties in New Jersey in 2017 and 2018, when the impacts of the Clean Truck Program revision will likely be the greatest. We also describe potential future trends in emissions and discuss limitations of the methods used here.

Background

The Port of New York & New Jersey (Port) is the third largest port in the U.S.A. In 2015 the Port moved over 3.6 million containers, an increase of more than 30% over 2005 levels (estimated from *14*). Port-related activities result in air pollution emissions including diesel exhaust and particulate matter, which are known to pose health risks (*1*). Heavy-duty diesel trucks emit about 40% of Port-related fine particulate matter ($PM_{2.5}$), with cargo handling equipment, locomotives, and marine vessels emitting the remainder (estimated from *15*). As the Port's activity has increased, surrounding communities have expressed concerns about the health impacts of air pollution from truck traffic and other port activities.

In 2009, the Port Authority of New York & New Jersey (PANYNJ) issued a Clean Air Strategy which outlined a number of measures to reduce air pollution from port activities. Strategies targeting port drayage trucks included operational improvements (congestion reduction, appointment system, and extended gate hours), truck technology improvements (encouraging or requiring the use of cleaner trucks), and land use planning strategies (development of warehouses and distribution centers closer to the Port.)

By 2011, funding for cleaner port drayage trucks had contributed to 429 replacements and 31 retrofits (*16*). The PANYNJ also committed to requiring drayage trucks to meet or exceed the U.S. Environment Protection Authority's (EPA) emissions standards for 2007 on-road heavy-duty vehicles or to use alternative fuel or hybrid technology starting January 1, 2017 (*17–19*). However, in light of concerns about the costs of the program and the potential for drayage truck shortages, in 2016 the PANYNJ revised (and significantly rolled back) the Clean Truck Program requirements. The revised Clean Truck Program requires that drayage trucks have an engine year of 1996 or newer or use alternative fuel or hybrid technology starting January 1, 2018 (*20*). Additionally, trucks registering to access the Port after March 1, 2016 must be 2007 engine year or newer. Trucks already registered that have an engine of 1996 or newer will continue to be permitted to enter the Port under the revised Clean Truck Program.

The revised Program will result in much smaller emissions reductions than the original Program. As of May 2016, 68% of the approximately 304,701 truck visits to the Port were completed by trucks that were older than model year 2007 (estimated from *21*). These trucks would be removed from service under the original Clean Truck Program but can remain in service under the revised Program. Only 5% of Port trucks are older than model year 1996 and will be removed from service under the revised Program (estimated from *21*). Consequently, the implementation of the revised Program will result in a far more modest shift in the truck fleet. These differences in the fleet's age profile are critical because prior to 2007 heavy-duty trucks were permitted to emit ten times as much $PM_{2.5}$ as their newer counterparts.

The purpose of this analysis is to quantify the predicted $PM_{2.5}$ impacts and the resultant changes in the risk of premature mortality in surrounding communities due to the revision of the Port's Clean Truck Program.

Data and Methods

A primary concern in communities adjacent to heavy-duty diesel truck traffic is exposure to fine particulates. We estimate the emissions of $PM_{2.5}$ from Port trucks under the original and revised Clean Truck Programs.

The scope of our analysis includes all container, auto, and warehouse terminal activities at the four busiest Port marine terminals, including New Jersey's Port Newark, Elizabeth Port Authority Marine Terminal, Port Jersey Port Authority Marine Terminal, and New York's Howland Hook Marine Terminal. The analysis examines drayage truck activities for those terminals, including on-terminal starts, idling, and driving, and off-terminal driving throughout the region.

We evaluate emissions from the four Port terminals and Port truck routes within eight New Jersey counties

(Bergen, Essex, Hudson, Middlesex, Morris, Passaic, Somerset, and Union). Using these location-specific emissions estimates, we then estimate the Census block-level change in annual average $PM_{2.5}$ concentrations that will result from the revision of the Clean Trucks Program. The health risks (all-cause premature mortality) of this change in concentrations are then evaluated for residents of the eight New Jersey counties. The analysis focuses on the first two years affected (2017 and 2018), which will exhibit the greatest effects. We also discuss the potential for ongoing future impacts.

Emissions

We first pair estimated emissions rates with estimated Port truck activities to determine overall hourly emissions on- and off-terminal under the original and revised Programs for each year, season, and time period.

Emissions Rates. We estimate Port truck $PM_{2.5}$ emissions rates from exhaust, brake wear, and tire wear for starts (as grams per start), short and extended idling (as grams per hour), and running (as grams per mile) using MOVES2014a. We use Essex County meteorological data from the New Jersey 2011 Periodic Emissions Inventory (dated June 2015) and MOVES2014a default fuel supply data (*22*). We evaluate emissions rates for four time periods corresponding to the travel demand model periods (AM: 6 a.m. to 9 a.m., MD: 9 a.m. to 3p.m., PM: 3 p.m. to 6 p.m., and NT: 6 p.m. to 6 a.m.) in January, April, July, and October (representing emission rates from January to March, April to June, July to September, and October to December, respectively). Running emissions are estimated for each road type and each speed category included in MOVES2014a.

To estimate the impacts of the revision of the Program, we estimate Program-specific vehicle ages under the revised and original Program in 2017 and 2018. We use the Port's truck visitation data from May 2016 to represent the age profile of the drayage fleet in 2016 (*21*). We then use the MOVES 2014 Age Distribution Projection Tool to estimate fleet turnover each year (*23*). Where certain older model years are prohibited under the Programs we redistribute those shares in proportion to the share of each model year that is permitted. Figure 1 summarizes the estimated truck age distribution for the original and revised Programs in 2017 and 2018. The implementation of the original Program would result in a largely updated fleet of trucks accessing Port terminals; in contrast, the changes reflected in the revised Program are far more modest in that they are delayed and expected to result in truck ages that are similar to the existing fleet.

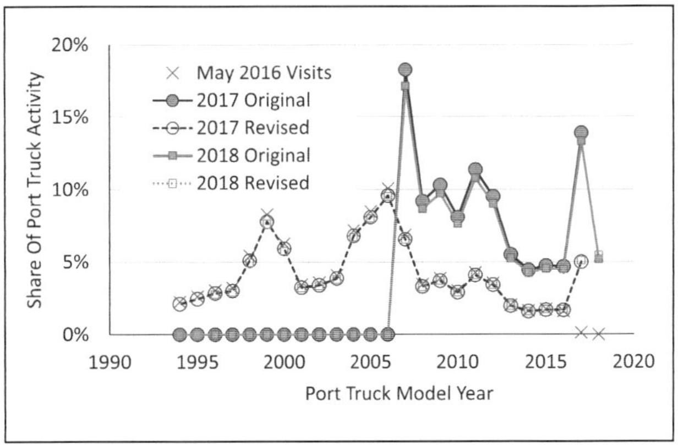

Figure 1. Port truck fleet assumptions for the original and revised Clean Truck Programs.

When estimating start, short idle, and running emissions, we assume that 95% of trucks are diesel combination short haul and 5% are diesel combination long haul (traveling greater than 200 miles) based on the approximate travel distances between all port terminals and their origins and destinations for trucks, estimate from (*24*). To estimate extended idling emissions we assume that all trucks are diesel combination long haul, as MOVES estimates extended idling rates for long-haul trucks only.

Truck Activity. We then estimate truck activity by terminal and route. The number of trucks and truck activity estimates are assumed to be the same under the original and revised Programs.

We estimate future volumes of container, motor vehicle, and bulk and general cargo at the Port. We assume linear growth in Port traffic based on trends from 1991 to 2015 (*14*, *25*), adjusting for on-dock rail (assuming 1.7 TEUs per lift). We then estimate the number of truck calls per corresponding commodity movement for each facility type based on 2013 truck call and freight activity data (*14*, *15*). Next, we allocate container, motor vehicle, and warehouse facility activities to each of the four terminals evaluated. Container allocations are based on 2000 and predicted 2020 values (*26*), adjusting for on-dock rail (*14*). We allocate auto and warehouse activity in proportion to the number of corresponding facilities at each terminal (*15*, *27*, *28*). We combine these estimates to predict truck calls for warehousing, auto, and container facilities at each terminal in 2017 and 2018. We estimate that there will be over 5 million truck calls at the four terminals in each year.

We estimate the share of hourly truck calls at each terminal for each season based on container traffic shares from 2005 to 2015 (*29*). We assume that on- and off-terminal Port-related truck activity occurs uniformly between 6 a.m. and 6 p.m. on 246 weekdays per year.

To estimate on-terminal activity for each truck call, we use the Port's estimates of 2013 on-terminal activities (*15*), with two minor deviations that resulted from a transcription error. Each of these errors leads to slightly underestimated emissions. First, we assume a truck speed of 15 mph at auto terminals instead of the Port's assumed 5 mph. Second, we assume that trucks at container terminals short idle for 0.43 hours instead of the Port's assumed 0.46 hours. Because auto-terminal truck calls are approximately 1.5% of all Port truck calls and on-terminal emissions are dwarfed by off-terminal emissions in terms of both magnitude and proximity to residents, the impact of these errors is likely to be small.

To evaluate off-terminal driving associated with each terminal we obtain the share of trucks traveling to and from locations across the country from the Port Truck Origin and Destination Survey (*24*). For the eleven counties nearest to the Port (which capture the bulk of Port truck traffic and include the eight study area counties), we use county-wide trip share estimates from (*24*). Additionally we disaggregate those trips to more spatially detailed locations based on 2015 heavy truck origins and destinations from the North Jersey Transportation Planning Authority's (NJTPA) travel demand model, the New Jersey Regional Transportation Model-Enhanced (NJRTM-E). While the travel analysis zones (TAZs) selected largely correspond to the terminal boundaries (particularly for the two busiest terminals), they likely also capture some non-Port trips.

We then create a travel network and estimate the routes between each terminal and its corresponding origins and destinations. The travel network is a combination of two datasets. For the NJTPA area we use the 2015 NJRTM-E loaded travel network. This road network includes road type, travel speeds for four time periods (morning, midday, afternoon, and night), and link type, which indicates routes on which trucks are prohibited. Outside of the NJTPA area we use the Federal Highway Administration's Freight Analysis Framework network, which includes highway travel speeds. We then estimate the quickest path between each terminal and its corresponding origins and destinations. We pair this route and origin–destination information with estimated terminal activity to arrive at estimated truck traffic volumes and speeds across the study area for the morning peak, midday, and evening peak periods. Figure 2 shows the share of all Port trucks traveling on each route for the morning peak period. Routes with no traffic are not shown. Midday and evening shares are similar.

Concentrations

Estimated emissions for each scenario (original and revised Clean Truck Program), season, and time period were used to estimate $PM_{2.5}$ concentrations in 2017 and 2018 that are expected to result from Port trucks in each scenario. The difference in concentrations for the two scenarios represents the change in $PM_{2.5}$ concentrations that is expected to result from the revision of the Clean Trucks Program.

The EPA's AERMOD steady state air dispersion model (version 15181) was used to estimate the annual average concentration (calculated as the average of daily concentration estimates) of primary emissions of $PM_{2.5}$ from Port-related truck activity on the regional transportation network and from on-terminal truck activities. AERMOD modeling procedures generally followed EPA $PM_{2.5}$ quantitative hotspot guidance (*30*) except as noted below. Each terminal was modeled as an area source with emissions arising from on-terminal truck activities assumed to be evenly dispersed throughout. This assumption is necessary because detailed data on truck activity within each terminal was not available. The perimeter of each area source aligns with the boundary of each terminal. Terminal boundaries are identified from Port terminal maps (obtained from the PANYNJ) and aerial imagery. Off-terminal truck activity is assigned to routes on the 2015 NJRTM-E roadway network. Each roadway link was defined as an area source whose length and width matched its dimensions. Five years of hourly meteorological data were obtained from New York State Department of Environmental Conservation (NYSDEC) for the Newark Airport monitoring station.

Several of our methods vary from the EPA guidance. Due to the large geographical area analyzed and the aim of estimating annual average concentrations rather than the 98th percentile 24-hour concentration required for conformity analysis, 24 days of data for each of the five years is used instead of the full five years required by EPA Hot Spot Guidance. Additionally, roadway segments with fewer than four daily truck trips were excluded from dispersion modeling to reduce computational time. Concentrations are modeled at receptors placed on a regular 100-m grid across the eight-county study area. Empirical Bayesian Kriging (implemented in ArcGIS version 10.3) was used to create a concentration raster with a 20-m grid cell resolution. The raster provides a smooth surface to aid in estimating population exposures and visualizing modeled results. Rowangould (*31*) provides a detailed discussion of these additional methods and their justification for use in modeling average concentrations from large transportation networks, as was done here.

Health Risks

The changes in annual average concentrations were then used to estimate the health risks associated with the Clean Truck Program revision at the Census-block level.

Figure 2. Estimated port truck route shares (AM peak period).

We restrict this analysis to the risk of premature mortality due to the robust body of health research tying premature mortality to $PM_{2.5}$ exposure (*32, 33*) and the limitation of the annual average $PM_{2.5}$ concentrations estimated in our dispersion analysis (which is consistent with long-term exposures underlying much of that research).

EPA's Environmental Benefits Mapping and Analysis Program-Community Edition (BenMAP-CE v. 1.1) was used to estimate risks of adult all-cause premature mortality in the eight-county study area due to changes in emissions in 2017 and 2018 alone. The BenMAP-CE modeling used 2010 Census block-level population and age data (from BenMAP's PopGrid) and built-in 2010 county-level health incidence data. Our $PM_{2.5}$ concentration estimates were aggregated to Census blocks as the mean of all 20-m grid located points within in each Census block. We use the health impact functions and pooling methods described in EPA's final regulatory assessment of the National Ambient Air Quality Standards (NAAQS) for particulate matter (*34*) with the inputs from the BenMAP-CE default $PM_{2.5}$ configuration and pooling setup applied to our New Jersey case.

Results and Discussion

Emissions

Emissions rates (per start, per hour, and per mile) for on-terminal activities are four to 18 times higher in 2017 and 2018 under the revised Clean Trucks Program when compared with the original Program. Per mile emissions rates for off-terminal driving are four to eight times higher in 2017 and 2018 under the revised Program when compared with the original Program.

Figure 3 shows the total emissions in 2017 and 2018 for on- and off-terminal activities under the original and revised Clean Truck Programs. There are 11 times more emissions from the four Port terminals under the revised Clean Truck Program in 2017 and 2018 when compared with the original Program. In the eight-county study area there are seven times more emissions along off-terminal truck routes under the revised Clean Truck Program in 2017 and 2018 when compared with the original Program. Total emissions from Port terminals and along truck routes in the study area increased by between 7.4 and 8 times.

Concentrations

Figure 4 shows the increases in $PM_{2.5}$ concentration that will result from the revised Clean Truck Program in the eight-county area in 2017. As expected, the greatest increases occur at Port terminals and along major truck routes accessing terminals. Increases in 2018 are similar (although slightly smaller) than in 2017. $PM_{2.5}$

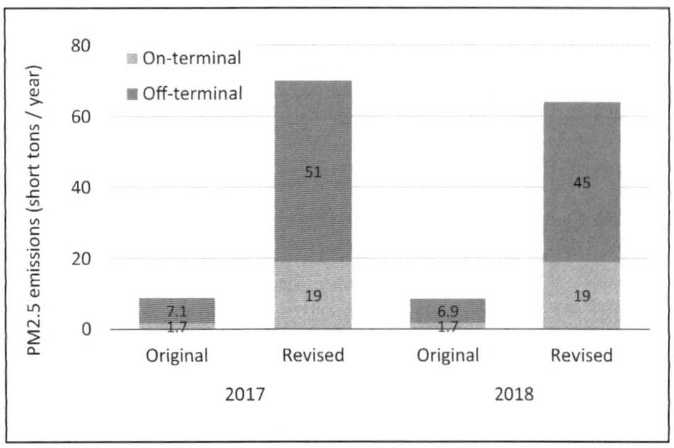

Figure 3. $PM_{2.5}$ emissions from port trucks under the original and revised Clean Truck Programs.

concentrations increase by up to 1 $\mu g/m^3$ in 2017 and 2018 in several areas under the revised Program when compared with the original Program. For reference, the annual $PM_{2.5}$ NAAQS is 12 μ/m^3, calculated as a three-year average.

Risks of Premature Mortality for Adults

Clearly conveying health risks is a critical step when evaluating the effects of a plan or program that affects air quality. Population-normalized risk estimates facilitate comparisons of the risks faced by individuals living in communities with different populations. When evaluating premature mortality, we use a one in a million risk of premature mortality over one year of exposure as a reference point. This analysis conservatively presents risks from one year of exposure due to the challenges of estimating exposures over the course of the program's lifetime; in reality the added risks accumulated over 10 or 20 years (or a 70-year lifetime) would be greater.

We estimate the added risks of all-cause premature mortality for adults across the eight-county study area that are expected to occur due to the revised Clean Truck Program. Premature mortality risk estimates based on two health impact studies (*32, 33*) are shown, consistent with EPA's final regulatory assessment of the NAAQS for particulate matter (*34*). These estimates are comparable to estimates based on functions derived by experts (calculated as in the EPA NAAQS regulatory assessment, results not shown.)

Results are presented as the increased risk of premature mortality (per million affected individuals) and are averaged at the municipality level (Table 1). We estimate that the increase in $PM_{2.5}$ emissions in 2017 alone will result in city-wide average increased risk of premature mortality of one to 10 in a million for adults living in several nearby communities, including the cities of Bayonne, Elizabeth, and Newark. In 2018 these risks are similar.

Note that these municipality-level averages combine areas of higher and lower risk. To identify the most at-risk residents (without averaging across areas of high and low risk) we also examine estimates at the population level for each county in the study area (Table 2). The increase in emissions in 2017 alone is estimated to result in approximately 400,000 to 700,000 adults in the region experiencing increased risks of premature mortality of at least one in a million, 5,000 to 40,000 residents are estimated to experience risks of premature mortality that increase by at least 10 in a million, and 600 to 3,000 adults are estimated to experience risks of premature mortality that increase by at least 25 in a million. Essex, Hudson, and Union counties exhibit the greatest increases in risks, with 24% to 39%, 36% to 52%, and 25% to 39% of their adult populations, respectively,

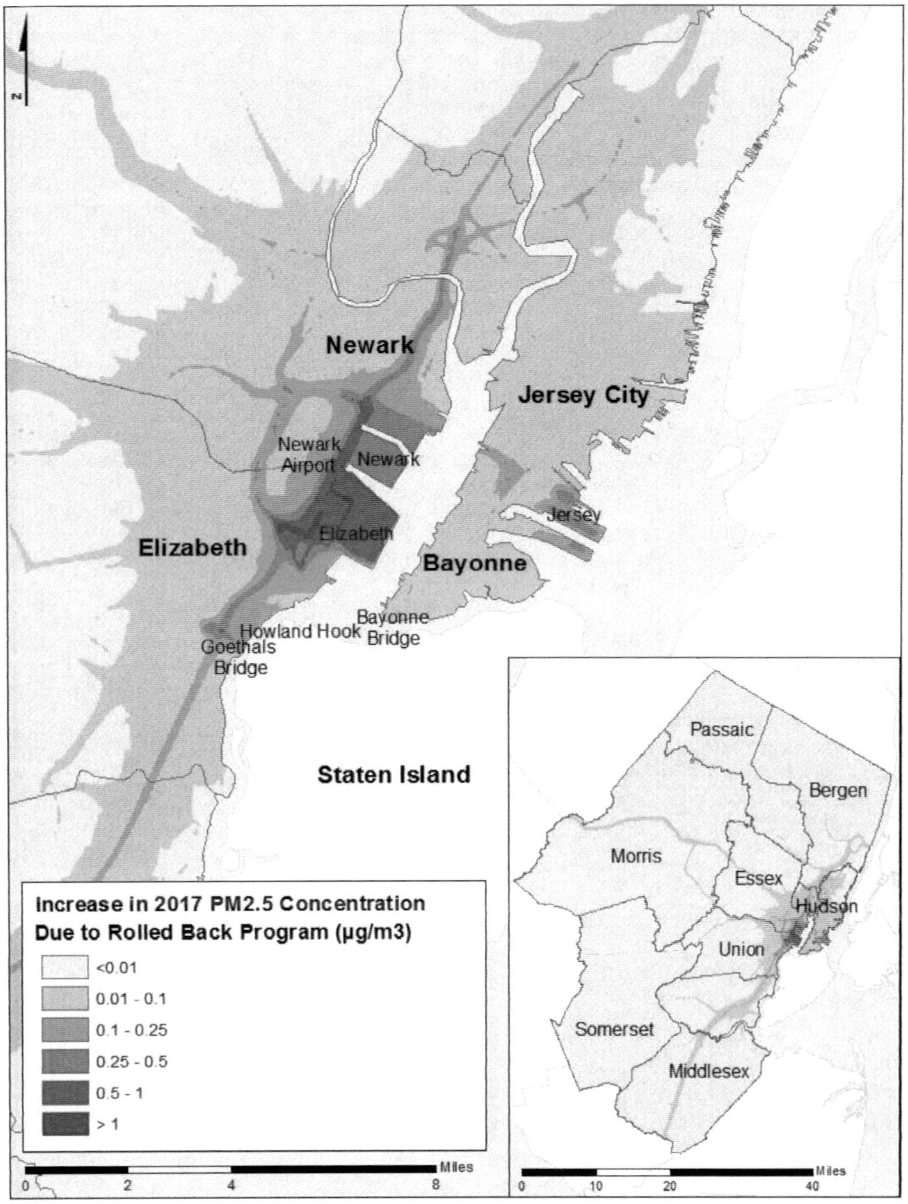

Figure 4. Increase in annual average $PM_{2.5}$ concentrations expected to result from the revised Clean Trucks Program in 2017.

experiencing increased risks of premature mortality of more than one in a million due to increases in $PM_{2.5}$ emissions in 2017 alone. Results in 2018 are similar.

These premature mortality estimates represent the increased risks of death from all causes. Based on (*32*), we estimate that approximately 67% to 92% of adult premature deaths due to increased $PM_{2.5}$ are attributable to increased risk of ischemic heart disease while approximately 8% to 16% of adult premature deaths are attributable to increased risk of lung cancer.

Future Trends

Increasing emissions from future growth in Port-related activity will be counteracted by improvements in truck per-mile and per-hour emissions rates as older vehicles

Table 1. Estimated City-wide Average Increase in the Risk of Premature Mortality Under the Revised Program (Compared with the Original Program) for the Nine Highest Risk Municipalities

Municipality	Number of affected adults[b] based on (26)	Number of affected adults based on (27)	2017 risk of premature mortality[a] based on (26) (per million affected adults[b])	2017 risk of premature mortality[a] based on (27) (per million affected adults[c])	2018 risk of premature mortality[a] based on (26) (per million affected adults[b])	2018 risk of premature mortality[a] based on (27) (per million affected adults[c])
Bayonne City	38,786	43,244	4.9 (3.3–6.5)	10 (5–14.9)	4.6 (3.1–6.1)	9.3 (4.6–13.9)
Elizabeth City	69,209	79,730	2.7 (1.8–3.6)	5.4 (2.7–8.1)	2.5 (1.7–3.3)	4.9 (2.4–7.4)
Newark City	149,272	173,009	1.8 (1.2–2.4)	3.7 (1.8–5.5)	1.7 (1.1–2.2)	3.3 (1.6–5.0)
Harrison Town	7,913	9,408	1.8 (1.2–2.3)	3.4 (1.7–5.1)	1.6 (1.1–2.1)	3 (1.5–4.5)
Hillside Township	13,115	14,484	1.4 (0.93–1.8)	2.8 (1.4–4.3)	1.3 (0.86–1.7)	2.4 (1.2–3.6)
Jersey City	139,568	168,007	1.4 (0.92–1.8)	2.6 (1.3–3.9)	1.3 (0.85–1.7)	2.6 (1.3–3.9)
Secaucus Town	10,679	11,839	1.3 (0.88–1.7)	2.6 (1.3–4.0)	1.2 (0.78–1.5)	2.4 (1.2–3.5)
East Newark Borough	1,250	1,492	1.3 (0.85–1.7)	2.4 (1.2–3.6)	1.1 (0.76–1.5)	2.1 (1.1–3.2)
Carteret Borough	12,670	14,219	1.1 (0.75–1.5)	2.2 (1.1–3.4)	1 (0.67–1.3)	2 (1.0–3.0)

[a]Average city-wide risk is shown; 95% confidence intervals are shown in parentheses.
[b]Krewski et al. (*32*) estimates include adults aged 30 and up.
[c]LePeule et al. (*33*) estimates include adults aged 25 and up.

are replaced with newer vehicles. The original Clean Truck Program would have greatly accelerated the turnover of the vehicle fleet. This analysis only captures impacts and risks that will occur due to changes in emissions in 2017 and 2018, which are expected to exhibit the largest emissions impacts and the health risks over the course of the revision's lifetime.

However, we anticipate that risks will continue to accrue beyond 2018. Although the computational resources required to evaluate $PM_{2.5}$ concentrations and health risks for several years into the future are prohibitive, we can approximate the trajectory of total emissions in order to provide qualitative insight about potential future trends in concentrations and health risks. We assess the total emissions (from on-terminal and off-terminal activities) in 2022, 2026, and 2030 using the same methods used for 2017 and 2018, although we caution that any uncertainties may be greater in future years.

Overall we find that the greatest impacts of the revised Clean Truck Program on total Port truck emissions occur initially and will likely lessen over the course of the next 15 years or more (Figure 5). We expect that concentration and health risk impacts would lessen on a similar timeline, although the shape of the trajectory of health risks would differ.

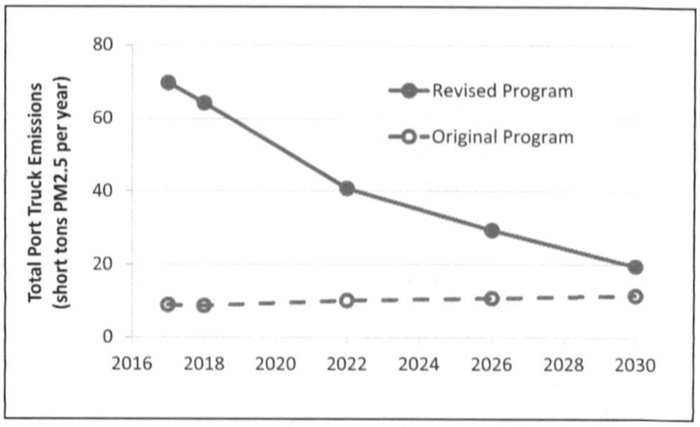

Figure 5. Future trends in port truck emissions under the original and revised Clean Truck Programs.

Limitations

As with any modeling exercise, the estimates presented here are constrained by computational power and available data.

Truck activity is estimated using the best publicly available information. The share of warehouse and auto activities at each terminal is less certain than estimates of the share of container activities, although this likely has

Table 2. Number of Adults at Risk of Premature Mortality due to Increased Exposure to $PM_{2.5}$ in 2017 Under the Revised Program (Compared with the Original Program)

County	>1 in a million risk[a,b] based on (*32*)	>1 in a million risk[a,c] based on (*33*)	>10 in a million risk[a,b] based on (*32*)	>10 in a million risk[a,c] based on (*33*)	>25 in a million risk[a,b] based on (*32*)	>25 in a million risk[a,c] based on (*33*)	Total adult population[b] for (*32*)	Total adult population[b] for (*33*)
Bergen	12,000 (7,300–17,000)	41,000 (14,000–68,000)	0 (0–0)	170 (0–1,500)	0 (0–0)	0 (0–0)	578,879	629,369
Essex	110,000 (82,000–130,000)	200,000 (130,000–260,000)	1,500 (1,300–1,900)	6,700 (1,500–22,000)	500 (0–1000)	1,300 (500–1,900)	458,208	511,024
Hudson	130,000 (97,000–160,000)	230,000 (150,000–270,000)	1,500 (210–5,800)	22,000 (2,300–47,000)	14 (0–73)	770 (15–5,400)	368,201	435,583
Middlesex	37,000 (22,000–49,000)	81,000 (41,000–110,000)	370 (16–790)	2,700 (410–6,200)	1.3 (0–1.3)	290 (1.3–640)	465,159	516,667
Morris	9,100 (6,200–12,000)	21,000 (10,000–32,000)	720 (0–750)	930 (0–1,300)	0 (0–0)	23 (0–760)	309,796	334,706
Passaic	510 (21–850)	4,300 (560–5,900)	0 (0–0)	0 (0–3.3)	0 (0–0)	0 (0–0)	288,664	321,928
Somerset	160 (16–210)	730 (200–2,500)	0 (0–0)	0 (0–0)	0 (0–0)	0 (0–0)	197,137	213,842
Union	82,000 (57,000–98,000)	14,0000 (93,000–170,000)	1,400 (300–3,000)	9,800 (1,600–22,000)	33 (0–260)	750 (36–3,000)	322,486	356,237
Total	380,000 (270,000–470,000)	710,000 (430,000–930,000)	5,400 (1,800–12,000)	42,000 (5,800–100,000)	550 (0–1,300)	3,100 (550–12,000)	2,988,531	3,319,355

[a]Number of adults experiencing increased risks of premature death are shown. 95% confidence intervals are shown in parentheses.
[b]Krewski et al. (*32*) estimates include adults aged 30 and up.
[c]LePeule et al. (*33*) estimates include adults aged 25 and up.

a small impact on our on-terminal emissions estimates because the majority of truck traffic is tied to container activities. If there are major divergences from current trends in on-dock rail and overall Port volumes in future years then our projections of truck volumes would fail to capture these differences. Our on-terminal activity estimates (speeds, idling time, time on-terminal) are based on the information presented in the Port's 2013 Inventory (*15*) although the data and the methods supporting those assumptions are sparse. Off-terminal travel speeds are estimated based on the NJRTM-E model outputs, which may differ from actual speeds in some areas.

Route estimates are subject to the errors built into the Port Truck Origin and Destination Survey (*24*) and the NJRTM-E due to their uncertainties and changes that may have occurred over time. We attempt to minimize the impacts of these potential errors by presenting health results in the aggregate rather than for each Census block across the region. Additionally, the last leg of each route was excluded, as centroid connectors do not represent actual route locations. We were also unable to account for truck idling and queuing activities that occur at off-terminal warehouses, chassis yards, or rail yards, due to a lack of information. The estimates of fleet emissions rates depend on the MOVES fleet projection tool's assumptions about fleet turnover rates, which may differ from actual fleet turnover for Port trucks.

The dispersion modeling is subject to the limitations of AERMOD's ability to predict pollution transport. Estimated $PM_{2.5}$ concentrations include only primary (but not secondary) formation of $PM_{2.5}$. Due to the computational demands of the spatially detailed dispersion modeling used, the $PM_{2.5}$ concentration outputs are constrained to annual average values (rather than short-term peak values), limiting the range of health impacts that can be assessed. Other health risks associated with short- and long-term $PM_{2.5}$ pollution (not considered in this analysis) include bronchitis, emergency room visits for asthma, lower and upper respiratory symptoms, nonfatal acute myocardial infarction, and hospital admissions for a variety of respiratory and cardiovascular concerns.

Additionally, health risk estimates are subject to the limitations of the assumed population and baseline mortality data. This analysis also assumes that diesel $PM_{2.5}$ emitted by Port trucks is the same as $PM_{2.5}$ emitted by other sources that drive the modeled health risk relationships although it may differ in composition. Additionally, emissions of other air pollutants such as nitrogen oxides (NO_x), hydrocarbons, and coarse particulate matter (PM_{10}) are not included in this analysis, although they likely contribute to increased local health risks.

Finally, due to computational limitations and uncertainties about future activities, this analysis was limited to the health impacts that are expected to occur due to $PM_{2.5}$ emissions in 2017 and 2018 alone. In reality, the revised Clean Truck Program is expected to result in elevated emissions that gradually taper off over a period of up to 15 years or more.

Conclusions

Clean truck policies for port drayage trucks can reduce emissions at port facilities as well as along truck routes outside of port boundaries. Estimating the pollution and health outcomes of clean truck policies can help to inform the design and implementation of pollution reduction plans for ports.

This analysis quantified predicted air quality impacts and one of the many health risks associated with scaling back the Port of New York & New Jersey's Clean Truck Program. Under the revised (rolled back) Clean Truck Program, most older (and higher emitting) Port trucks will still be allowed to access the Port. As a result, the revised Clean Truck Program will allow increased pollution from $PM_{2.5}$ near the Port of New York & New Jersey when compared with the original Clean Trucks Program.

We find that Port truck $PM_{2.5}$ emissions will increase by seven to eight fold in 2017 and 2018 which will lead to increases of $PM_{2.5}$ concentrations by up to 1 μ/m^3 at some locations. In some areas, the risks of death will increase by over 25 in a million due to increases of emissions in 2017 alone. This analysis focuses on 2017 and 2018, the first two years of the revised Program, when the impacts of the revised Program will likely be the greatest. Elevated pollution levels and health risks will likely continue to accrue for up to 15 years or more after the Program has been revised.

Addressing the uncertainties described in the limitations section could improve the precision of the estimates. Additionally, these findings are limited in scope due to resource constraints. The magnitude of estimated effects of the revised Program would be greater if this study had included a broader geographic region, a longer period of time, and a wider range of truck pollutants, exposure pathways, and health outcomes. Despite the limited scope of this analysis, we estimate that the risks of serious adverse health outcomes will increase as a result of the revision of the Clean Truck Program.

Acknowledgments

This analysis was conducted on behalf of the Coalition for Healthy Ports. The authors also gratefully acknowledge the NJTPA for providing NJRTM-E data files and NYSDEC for providing AERMET files.

Author Contributions

The authors confirm contribution to the paper as follows: study conception and design: Dana Rowangould, Gregory Rowangould; data collection: Dana Rowangould; analysis and interpretation of results: Dana Rowangould, Gregory Rowangould, Debbie Niemeier; draft manuscript preparation: Dana Rowangould, Gregory Rowangould. All authors reviewed the results and approved the final version of the manuscript.

References

1. Bailey, D., and G. Solomon. Pollution Prevention at Ports: Clearing the Air. *Environmental Impact Assessment Review*, Vol. 24, No. 7–8, 2004, pp. 749–774.
2. Ault, A. P., M. J. Moore, H. Furutani, and K. A. Prather. Impact of Emissions from the Los Angeles Port Region on San Diego Air Quality during Regional Transport Events. *Environmental Science & Technology*, Vol. 43, No. 10, 2009, pp. 3500–3506.
3. Karner, A. A., D. S. Eisinger, S. Bai, and D. A. Niemeier. Mitigating Diesel Truck Impacts in Environmental Justice Communities. *Transportation Research Record: Journal of the Transportation Research Board*, 2009. 2125: 1–8.
4. Houston, D., M. Krudysz, and A. Winer. Diesel Truck Traffic in Low-Income and Minority Communities Adjacent to Ports: Environmental Justice Implications of Near-Roadway Land Use Conflicts. *Transportation Research Record: Journal of the Transportation Research Board*, 2008. 2067: 38–46.
5. Wu, J., D. Houston, F. Lurmann, P. Ong, and A. Winer. Exposure of PM2.5 and EC from Diesel and Gasoline Vehicles in Communities Near the Ports of Los Angeles and Long Beach, California. *Atmospheric Environment*, Vol. 43, No. 12, 2009, pp. 1962–1971.
6. Kozawa, K. H., S. A. Fruin, and A. M. Winer. Near-road Air Pollution Impacts of Goods Movement in Communities Adjacent to the Ports of Los Angeles and Long Beach. *Atmospheric Environment*, Vol. 43, No. 18, 2009, pp. 2960–2970.
7. Bishop, G. A., B. G. Schuchmann, D. H. Stedman, and D. R. Lawson. Emission Changes Resulting from the San Pedro Bay, California Ports Truck Retirement Program. *Environmental Science & Technology*, Vol. 46, No. 1, 2012, pp. 551–558.
8. Dallmann, T. R., R. A. Harley, and T. W. Kirchstetter. Effects of Diesel Particle Filter Retrofits and Accelerated Fleet Turnover on Drayage Truck Emissions at the Port of Oakland. *Environmental Science & Technology*, Vol. 45, No. 24, 2011, pp. 10773–10779.
9. Kuwayama, T., J. R. Schwartz, R. A. Harley, and M. J. Kleeman. Particulate Matter Emissions Reductions due to Adoption of Clean Diesel Technology at a Major Shipping Port. *Aerosol Science and Technology*, Vol. 47, No. 1, 2013, pp. 29–36.
10. Lee, G., S. You, S. G. Ritchie, J.-D. Saphores, R. Jayakrishnan, and O. Ogunseitan. Assessing Air Quality and Health Benefits of the Clean Truck Program in the Alameda Corridor, CA. *Transportation Research Part A: Policy and Practice*, Vol. 46, No. 8, 2012, pp. 1177–1193.
11. Norsworthy, M., and E. Craft. Emissions Reduction Analysis of Voluntary Clean Truck Programs at US Ports. *Transportation Research Part D: Transport and Environment*, Vol. 22, 2013, pp. 23–27.
12. Beall, S. *California SB-1 Transportation Funding (2017 - 2018)*. 2017.
13. Port of Long Beach and The Port of Los Angeles. *San Pedro Bay Ports Clean Air Action Plan 2017 Draft Final July 2017*, 2017.
14. Port Authority of New York / New Jersey. *2015 Trade Statistics The Port of New York and New Jersey 2005 - 2015*, 2016.
15. Starcrest Consulting Group, LLC. *The Port of New York and New Jersey Port Commerce Department 2013 Multi-Facility Emissions Inventory*. Port Authority of New York and New Jersey, 2015.
16. Bonney, J. *NY-NJ Port's Truck Replacement Program Shifts into Gear*. The JOC Group, Inc., Newark, NJ, 2015.
17. Port Authority of New York / New Jersey. *Naming Rules and Regulations Applying At Port Authority Marine Terminals And Rates and Charges Applicable For the Use of Public Areas At Port Authority Marine Terminals*. New York, NY, 2014.
18. Port Authority of New York / New Jersey. *Naming Rules and Regulations Applying At Port Authority Marine Terminals And Rates and Charges Applicable For the Use of Public Areas At Port Authority Marine Terminals*. New York, NY, November 2015.
19. Port Authority of New York / New Jersey. *Naming Rules and Regulations Applying At Port Authority Marine Terminals And Rates and Charges Applicable For the Use of Public Areas At Port Authority Marine Terminals*. New York, NY, September 2015.
20. Port Authority of New York / New Jersey. *Naming Rules and Regulations Applying At Port Authority Marine Terminals And Rates and Charges Applicable For the Use of Public Areas At Port Authority Marine Terminals*. New York, NY, 2016.
21. Port Authority of New York / New Jersey. *PortTruckPass Comprehensive Report*. May 2016.
22. U.S. Environmental Protection Agency. *MOVES Technical Guidance: Using MOVES to Prepare Emission Inventories for State Implementation Plans and Transportation Conformity*. 2015.
23. U.S. Environmental Protection Agency. *Age Distribution Projection Tool for MOVES 2014*. 2014.
24. Vollmer, T. A. Eng-Wong, Stump/Hausman, New Jersey Institute of Technology and Stevens Institute of Technology. *Port Authority Marine Container Terminals Truck Origin-Destination Survey 2005 (Draft Report, Revised 2/27/2006)*. The Port Authority of New York and New Jersey, 2006.
25. Port Authority of New York / New Jersey. *2011 Trade Statistics The Port of New York and New Jersey*. 2012.
26. Halcrow, Gannett Fleming, MDS Transmodal, Duncan Maritime, Moffatt & Nichol Engineers, Zetlin Strategic

Communications and Hirani Engineering. *CPIP Consortium Port of New York & New Jersey Comprehensive Port Improvement Plan*. 2005.
27. Port Authority of New York / New Jersey. *Port of New York and New Jersey - Distribution and Warehousing Centers*. http://www.panynj.gov/port/distribution-warehousing.html. Accessed 2016.
28. Port Authority of New York / New Jersey. *Port of New York and New Jersey - Automotive Logistics*. http://www.panynj.gov/port/vehicle-shipping-processing.html. Accessed 2016.
29. Port Authority of New York / New Jersey. *Loaded Containers in TEUs and Total ExpressRail Lifts by Month*. http://www.panynj.gov/port/monthly-loaded-containers.html. Accessed 2016.
30. U.S. Environmental Protection Agency. *Transportation Conformity Guidance for Quantitative Hot-spot Analyses in PM2.5 and PM10 Nonattainment and Maintenance Areas*. 2015.
31. Rowangould, G. M. A new approach for evaluating regional exposure to particulate matter emissions from motor vehicles. *Transportation Research Part D: Transport and Environment*, Vol. 34, 2015, pp. 307–317.
32. Krewski, D., M. Jerrett, R. T. Burnett, R. Ma, E. Hughes, Y. Shi, M. C. Turner, C. A. Pope 3rd, G. Thurston, E. E. Calle, M. J. Thun, B. Beckerman, P. DeLuca, N. Finkelstein, K. Ito, D. K. Moore, K. B. Newbold, T. Ramsay, Z. Ross, H. Shin, and B. Tempalski. *Extended Follow-up and Spatial Analysis of the American Cancer Society Study Linking Particulate Air Pollution and Mortality*. Publication 1041-5505 (Print). Health Effects Institute, Boston, MA, 2009.
33. Lepeule, J., F. Laden, D. Dockery, and J. Schwartz. Chronic Exposure to Fine Particles and Mortality: An Extended Follow-up of the Harvard Six Cities Study from 1974 to 2009. *Environ Health Perspectives*, Vol. 120, No. 7, 2012, pp. 965–970.
34. U.S. Environmental Protection Agency. *Regulatory Impact Analysis for the Final Revisions to the National Ambient Air Quality Standards for Particulate Matter*. Research Triangle Park, NC, 2012.

The Standing Committee on Marine Environment (AW030) peer-reviewed this paper (18-06383).

Article

TRR
JOURNAL OF THE TRANSPORTATION RESEARCH BOARD

Transportation Research Record
2018, Vol. 2672(11) 65–72

DOI: 10.1177/0361198118787388
journals.sagepub.com/home/trr
SAGE

A Zero-Inflated Negative Binomial Regression Model to Evaluate Ship Sinking Accident Mortalities

Tian Chai[1,2], De-qi Xiong[1], and Jinxian Weng[3]

Abstract

Sinking accidents are a seafarer's nightmare. Using 10 years' of worldwide sinking accident data, this study aims to develop a mortality count model to evaluate the human life loss resulting from sinking accidents using zero-inflated negative binomial regression approaches. The model results show that the increase of the expected human life loss is the largest when a ship suffers a precedent accident of capsizing, followed by fire/explosion or collisions. Lower human life loss is associated with contact and machinery/hull damage accidents. Consistent with our expectation, cruise ships involved in sinking accidents usually suffer more human life loss than non-cruise ships and there is be a bigger mortality count for sinking accidents that occur far away from the coastal area/harbor/port. Fatalities can be less when the ship is moored or docked. The results of this study are beneficial for policy-makers in proposing efficient strategies to reduce sinking accident mortalities.

The shipping industry has witnessed a fast growth in the last three decades because of the significant increase of transportation demands. Shipping movements are operated in a complex and high-risk environment, and many shipping accidents occur at sea, as well as in restricted waters (*1, 2*). More and more shipping liner companies have an increasing interest in using large-sized ships in the foreseeable future because of their scale advantages. However, a large ship (e.g., a cruise ship) has reduced maneuverability, which ultimately is an increased risk (*3*). As a bigger size brings a corresponding increase in cargo and passengers, it may lead to catastrophic consequences with regards to human life loss. Therefore, sinking accidents are a frightening consideration for seafarers and any incidents are likely to be subjected to much public criticism.

It is incumbent for policy-makers to implement efficient navigational safety strategies with the objectives of reducing the likelihood of sinking accidents and rescuing more people from these accidents. Considering the limited available resources and budgets, policy-makers have to determine the prioritization of safety strategies. This can be achieved with the help of a comprehensive understanding of the contributory factors affecting the mortalities that have resulted in sinking accidents.

So far, numerous studies have been carried out for the analysis of shipping accidents. The major focus has been placed on fishing accidents in these studies, such as the investigation of contributory factors, and their effects on the fishing accident probability and consequence (*4–8*). However, it should be noted that the fishing accident results were only applicable to fishing vessels. Considering all ship types, some other researchers (*9, 10*) also investigated the relationship between the contributory factors and the risk of shipping accidents. Nevertheless, their results may still be biased because the data sources consider all possible accident types. In general, the contributory factors may present varying effects on the accident mortality for different accident types. In addition, some accident types may be related to each other. For example, a ship may suffer a collision, fire/explosion, contact, grounding and/or capsizing before the occurrence of ship sinking (*11*). It was found from the historical data that the majority of shipping accidents may result in no mortality because of efficient emergency rescue plans and other unknown reasons. Zero-inflated models have been considered as a useful approach to

[1]Environmental Science and Engineering College, Dalian Maritime University, Dalian, China
[2]Navigation Institute, Jimei University, Xiamen Fujian, China
[3]College of Transport and Communications, Shanghai Maritime University, Shanghai, China

Corresponding Author:
Address correspondence to Jinxian Weng: jxweng@shmtu.edu.cn

model the count data (e.g., the ship mortality) with an excessive number of zero outcomes (*12–14*). Considering the large numbers of human life loss resulting from sinking accidents, this study aims to model the human life loss caused by shipping sinking accidents using zero-inflated negative binomial regression techniques.

Literature Review

To date, many researchers (*4–7*) have focused their attention on fishing vessel accidents. For example, Jin et al. examined the determinants of the total losses and number of fatal and non-fatal crew injuries resulting from commercial fishing vessel accidents (*4*). It was found that the probability of a total loss of the vessel was the highest for a capsizing, followed by a sinking accident. In addition, fire/explosions and capsizing were expected to incur the greatest number of crew fatalities. In addition, some researchers also examined ship accidents involved tankers (*15*), passenger ships (*16*) and cellular type containerships (*17*).

It should be pointed out that the results from the above studies were only applicable for a specific ship type. In other words, these studies cannot examine the effect of the ship type on the consequence of the shipping accident. On the other hand, many other studies have been conducted to investigate the relationship between the contributory factors and shipping accident risks considering all ship types. For example, some researchers analyzed shipping accidents and proposed many navigation safety enhancement strategies in the Istanbul Strait (*18, 19*). Debnath and Chin explored the influencing factors of shipping collision risks in the Singapore Port waters (*20*). Weng et al. examined the effects of time and traffic directions on shipping accident frequency in the Singapore Strait (*21*).

Nevertheless, the results from these studies may still be biased even though all ship types were taken into account. This is because the data sources used for the analysis were subject to specific water areas (e.g., the Istanbul Strait). In addition, the majority of existing studies were concerned with the ship accident occurrence probability and corresponding risk mitigation strategies. However, the general public have great concerns about sinking accidents because of the possible serious accident consequence (i.e., a large amount of human life loss). Unfortunately, there has not been a great deal of literature published on the analysis of mortality resulting from sinking accidents and the corresponding contributory factors.

Data

The worldwide shipping sinking accidents that occurred between the January of 2001 and February of 2011 were obtained from the shipping accident database managed by the Lloyd's List Intelligence Company. From the shipping accident database, we extracted all sinking accident related information on the: (i) ship type; (ii) precedent accident; (iii) operating conditions; (iv) accident location; and (v) the mortalities (the number of passengers and crew members who died or were declared missing in the accident). The operating conditions include the detailed information on the weather conditions, the operating time and dock condition. In this study, the operating time was divided into two categories: daytime and night-time periods. Hereafter, the daytime period is defined as the period from the local time of sunrise to the time of sunset. It should be pointed out that the times of sunrise and sunset vary with the time of year. The accident location includes "far away from the coastal area/harbor/port" and "near the coastal area/harbor/port". In this study, ships are categorized into three groups: (i) passenger ships; (ii) ferries/roll on-roll off (ro-ro) ships and (iii) other ship types (e.g., container carrier, cargo ship). The precedent accidents include collision, fire/explosion, machinery/hull damage/failure (e.g., lost rudder, fouled propeller), contact, grounding and capsizing. Hereafter, the collision is defined as a situation that the ship struck or was struck by another ship on the water surface. Contact refers to a situation in which the ship struck any fixed or floating objects other than those included under collision or grounding. A grounding situation refers to a situation in which the ship is in contact with the sea bottom or a bottom obstacle, such as a struck object on the sea floor, or struck or touched the bottom. Capsizing refers to a situation in which a ship is turned on its side or it is upside down.

From the shipping accident database, a total of 1871 records of ship accidents which occurred at 33 major worldwide water areas between the January of 2001 and the February of 2011 were collected in this study. Table 1 presents the variables and their descriptive statistics. The mean statistics reveal that 14.1%, 7.8%, 11.1%, 1.4%, 4.6% and 16.5% ships suffer a collision, fire/explosion, machinery/hull damage, contact, grounding and capsizing before the occurrence of ship sinking. Approximately 1.0% of ships involved in sinking accidents are cruise ships. The majority of sinking accidents occurred in good weather conditions (89.5%), and 44.8% of accidents occurred during the night-time period. Furthermore, 4.8% of sinking accidents occurred when the ship was docked or moored. In addition, a relatively small proportion of sinking accidents occurred far away from the coastal area/harbor/ports (15.8%). The loss of human life resulting from the shipping accidents studied ranges from 0 to 1800, with an average value of 5.43 persons and a standard deviation of 54.50 persons. The collected data show that the majority of accidents (69.7%) cause no mortalities (i.e., a zero count).

Table 1. Variable Definitions and Corresponding Statistics

Variables	Attributes	Mean	Standard deviation	Minimum	Maximum
Dependent variable					
Mortality	Number of people who were killed or missing in a sinking accident	5.432	54.50	0	1800
Independent variable					
Ship type	2 for passenger ships, 1 for ro-ros and ferries, 0 otherwise	0.042	0.244	0	2
Precedent accident					
Collision	1 if a collision is occurred before the occurrence of sinking, 0 otherwise	0.141	0.348	0	1
Fire/explosion	1 if a fire/explosion occurred before the occurrence of sinking, 0 otherwise	0.078	0.268	0	1
Machinery/hull damage	1 if machinery/hull damage occurs before the occurrence of sinking, 0 otherwise	0.111	0.314	0	1
Contact	1 if contact occurs, 0 otherwise	0.014	0.119	0	1
Grounding	1 if grounding occurs, 0 otherwise	0.046	0.209	0	1
Capsizing	1 if capsizing occurs, 0 otherwise	0.165	0.371	0	1
Operating conditions					
Weather conditions	1 if the accident occurred under adverse weather conditions, 0 otherwise	0.105	0.306	0	1
Operating time	1 for the night-time period, 0 otherwise	0.448	0.497	0	1
Docking condition	1 if the ship was docked or moored, 0 otherwise (e.g., underway)	0.048	0.214	0	1
Accident location	1 if the accident occurred far away from the coastal area/harbor/ports, 0 otherwise	0.158	0.365	0	1

Model Formulation

In traffic safety analysis, the count data (e.g., the human life loss) often has an excessive number of zero outcomes. To deal with the problem of excess zeros, one of the commonly used methods is zero-inflated distribution. A zero-inflated distribution is actually a mixture of two distributions including a delta distribution on zero ("perfect state") and a distribution on the non-negative integers ("imperfect state"). In general, a data record is in the perfect state with a probability p and in the imperfect state with probability $1-p$. If the data record is in the perfect state, it takes only the value zero. If the record is in the imperfect state, it follows a distribution on non-negative integers (including the value of zero). Lambert proposed a zero-inflated Poisson (ZIP) regression model in which the probability p was related to covariates using a logistic regression model, and a log-linear regression model was developed to relate the Poisson mean to covariates in the imperfect state (*22*). As an alternative to ZIP regression, one may consider zero-inflated negative binomial (ZINB) regression if the count data continues to suggest additional over-dispersion (*23*). In a ZINB regression model, the negative binomial model is used for the imperfect state. As mentioned above, Table 1 indicates that the standard deviation of mortalities is substantially higher than the mean. Therefore, the ZINB regression approach is more appropriate to model the number of mortalities resulting from sinking accidents in this study.

In a ZINB model, a natural set of stating values for the covariates is provided by the Probit or Logit and independent Poisson or negative binomial estimates. To determine the probability of p for the perfect state, another efficient method is to use the Poisson or negative binomial estimates which are multiplied by a value of τ. More specifically, the probability function of a ZINB regression model can be expressed by:

$$f(y_i|\boldsymbol{X}_i,\boldsymbol{\beta},\tau,\alpha)=\begin{cases}p_i+(1-p_i)q(0|\mu_i,\alpha), & \text{if } y_i=0\\ (1-p_i)q(y_i|\mu_i,\alpha), & \text{if } y_i>0\end{cases} \quad (1)$$

$$q(y_i|\mu_i,\alpha)=\frac{(\Gamma(y_i+\alpha^{-1})/y_i!/\Gamma(\alpha^{-1}))(\alpha^{-1}/(\alpha^{-1}+\mu_i))^{\alpha^{-1}}(\mu_i/(\alpha^{-1}+\mu_i))^{y_i}}{1-(1+\alpha\mu_i)^{-\alpha^{-1}}} \quad (2)$$

$$\ln(\mu_i)=\beta_0+\beta_1x_{i1}+\beta_2x_{i2}+\cdots+\beta_kx_{ki}=\boldsymbol{X}_i\boldsymbol{\beta} \quad (3)$$

$$p_i=p_i(y_i=0|\boldsymbol{X}_i,\boldsymbol{\beta},\tau)=\Phi(\tau\boldsymbol{X}_i\boldsymbol{\beta}) \quad (4)$$

In which $\boldsymbol{X}_i$ is the vector of the explanatory variables for the negative binomial regression model, $\boldsymbol{\beta}$ is the vector of coefficients, p_i is the probability of perfect state, α is the dispersion parameter, and Φ denotes the cumulative

Table 2. Statistical Comparison among Different Regression Models

Model performance	Poisson regression	Negative binomial regression	Negative binomial regression with normal heterogeneity	Negative binomial hurdle model	ZIP regression	ZINB regression
DIC (smaller is better)	15292	5477	5476	5392	7936	5368
AIC (smaller is better)	15295	5480	5479	5393	7939	5371
BIC (smaller is better)	15373	5558	5557	5472	8017	5449

Note: ZIP = zero-inflated Poisson; ZINB = zero-inflated negative binomial; DIC = deviance information criterion; AIC = Akaike information criterion; BIC = Bayesian information criterion.

distribution function of a standard normal random variable.

We can determine the estimates of $\boldsymbol{\beta}$, τ and α by maximizing the log-likelihood function $\ln L(\boldsymbol{\beta},\tau,\alpha|y,X) = \sum_{i=1}^{N} \ln f(y_i|X_i,\boldsymbol{\beta},\tau,\alpha)$ with respect to $\boldsymbol{\beta}$, τ and α. However, note that it is still unknown which state the zero-valued observations belong to. This implies that the procedures for estimating the coefficients of the generalized linear models and generalized additive models may not be applicable for the ZINB regression models. We can use the EM algorithm that is similar to the quasi-Newton methods to determine the maximum likelihood estimates. In the EM algorithm, a new random variable Z is introduced. Note that the random variable Z is only partially observable. If y is strictly greater than zero, the value of Z is known (i.e., zero) while it is unknown otherwise. More specifically, the EM algorithm at the jth iteration includes the following three steps:

(1) E-step: Compute

$$Z_i^j = E\left[Z_i|y_i,\boldsymbol{\beta}^{j-1},\tau^{j-1},\alpha^{j-1}\right] = \begin{cases} \frac{p_i^{j-1}}{p_i^{j-1}+(1-p_i^{j-1})q(0|\mu_i^{j-1},\alpha^{j-1})}, & \text{if } y_i = 0 \\ 0, & \text{if } y_i>0 \end{cases}$$

(2) M-step for $\boldsymbol{\beta}$ and α: Determine the estimates $\boldsymbol{\beta}^j$ and α^j by fitting the negative binomial regression model using the weights $(1\text{-}Z_i^j)$ and the response variable y_i.

(3) M-step for τ: Determine the estimate τ^j by fitting the Probit regression model using the $\boldsymbol{\beta}^j$ and response variable Z_i^j.

The details of the EM algorithm can be found in Lambert (*22*) and Hall (*24*). Minami et al. also explained how to implement the EM algorithm for the ZINB regression model (*23*).

Results and Discussions

Model Results

The zero-inflated regression procedure in the Limdep software (Version 9, Econometric Software Inc., NY, USA) was performed to calibrate the proposed ZINB regression model using the collected 1871 sets of sinking accident data. We also developed another three regression models, namely the Poisson regression model, the negative binomial regression model and the ZIP regression model to demonstrate that the ZINB model provides the best fit. When several models are available, one can compare the model performance using the most regularly used measures such as the Akaike information criterion (*AIC*), the Bayesian information criterion (*BIC*) and the Deviance information criterion (*DIC*). The *AIC* penalizes a model with a larger number of parameters and is defined as $AIC = -2\ln L + 2p$, in which $\ln L$ represents the fitted log-likelihood and p is the number of parameters for the model. The *BIC* penalizes a model with a larger number of parameters and a larger sample size and is defined as $BIC = -2\ln L + p\ \ln(N)$, in which N is the sample size. In general, lower values of the *AIC* and *BIC* statistics are preferable. Table 2 gives the statistical comparison results of different regression models. It can be seen from the table that the ZINB regression model provides the best fit to the sinking accident data because of the smallest *AIC*, *BIC* and *DIC* statistics. These results confirm the appropriateness of using the ZINB regression technique to model the loss of human life resulting from sinking accidents.

Table 3 shows the statistical analysis results for the mortality count model. From the table, it can be seen that the coefficient for fire/explosion is positive and statistically significant at a 0.05 level, indicating that there are more deaths and missing people if the ship suffers a fire/explosion before sinking into the water. The overdispersion parameter α in the model is positive and statistically significant at a 0.001 confidence level, suggesting that the data in the imperfect state is still overdispersed. This further confirms that the ZINB regression model outperforms the ZIP regression model. The model results also show that the mortality count in sinking accidents is significantly influenced (at a significance level of 0.001) by the following explanatory variables: grounding, capsizing, weather conditions, docking conditions and accident location. The coefficients associated with the collision, machinery/hull damage and operating

Table 3. ZINB Regression Model Results for Sinking Accidents

Variables	Coefficients	Standard errors	z-statistics
Negative binomial regression part			
Constant	0.487	0.065	7.42*
Ship type	2.638	0.289	9.11*
Collision	0.354	0.117	3.02**
Fire/explosion	0.372	0.145	2.56**
Machinery/hull damage	−0.276	0.121	−2.28**
Contact	−0.418	0.253	−1.65
Grounding	−0.603	0.162	−3.72*
Capsizing	0.468	0.116	4.04*
Weather conditions	0.535	0.147	3.63*
Docking condition	−1.253	0.256	−4.88*
Operation time	0.164	0.078	2.10**
Accident location	0.377	0.113	3.32*
α	3.044	0.084	36.2*
Probit regression part			
τ	−0.555	0.075	−7.37*

*p < .001. **p < .05.

time are statistically significant at a significance level of 0.05.

According to Table 3, it can be found that the occurrence of precedent events including collision, fire/explosion and capsizing will increase the number of deaths and missing people, which is expected to be greater when the ship suffers a precedent event including collision, fire/explosion and capsizing. However, a lower mortality count is associated with a precedent event such as machinery/hull damage, contact, grounding, and so on. In addition, human life loss will increase if the sinking accident occurs under adverse weather conditions (e.g., strong winds, waves). The negative sign for the docking conditions indicates that the loss of human life resulting from sinking accidents is less when the ship involved is docking or moored. The positive sign for the ship type shows that cruise ships are associated with a bigger number of mortalities in sinking accidents.

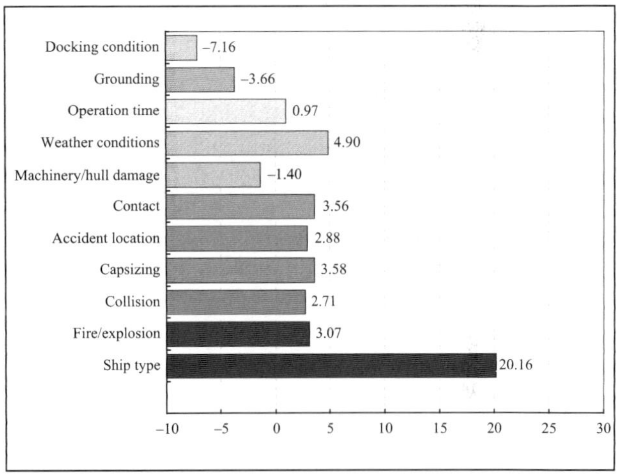

Figure 1. Marginal effects of contributory factors on sinking accident mortalities.

Marginal Effects

Although the signs of the estimated coefficients for the ZINB regression model could provide information on whether changes in given explanatory variables increase or decrease the loss of human life resulting from sinking accidents, they cannot provide further information on the extent to how much the loss of human life is changed. To gain further insight into the effects of the contributory factors on sinking accident mortalities, we can estimate the marginal effect of the explanatory variable x_i by:

$$\frac{\partial E(y|\boldsymbol{X}, \boldsymbol{\beta}, \tau, \alpha)}{\partial x_i} = \frac{\partial\{[1 - \Phi(\tau u)]u\}}{\partial x_i} = \beta_i u[1 - \Phi(\tau u) - \tau\phi(\tau u)] \tag{5}$$

In which u is the expected value of the estimates from the negative binomial model, ϕ is the standard normal density function, and β_i is the coefficient of the explanatory variable x_i used in the ZINB regression model. The marginal effects of explanatory variables on the mortality count resulting from sinking accidents are shown in Figure 1.

Effects of Ship Type. As expected, the number of mortalities resulting from shipping accidents is significantly affected by the ship type. More specifically, on average, the loss of human life for passenger ships is larger than for ferries/ro-ro ships by 20.16, as shown in Figure 1.

This result could possibly be because passenger ships capsize very quickly (*25*). Clearly, a quick capsizing event may result in a high number of fatalities. Another possible reason might be that the efficiency of an emergency evacuation could be low on a ship carrying many passengers (*26*). In particular, many people finding themselves in the confines of a ship at sea can also expect to experience signs of panic when a fire is present. This result provides evidence to support the argument of House that passenger ships have the potential for incurring high casualty rates in the event of major shipping accidents (*27*). To enhance the passenger ship safety level, all passenger ships are required to be fitted with emergency towing arrangements for fore and aft, pre-rigged and capable of easy deployment.

Effects of Precedent Accident. Figure 1 shows that the loss of human life is the highest for the precedent accident of capsizing, followed by fire/explosion and collision. The loss of human life will dramatically increase by 3.77 if a ship suffers a capsizing before eventually sinking into the sea. It is found from Figure 1 that the mortality count for the ship suffering a precedent accident of fire/explosion is on average 2.66 higher than for a ship suffering no fire/explosion before sinking into the sea. This might be explained by the consideration that sudden and overwhelming events provide little evacuation time for the passengers or crew members (*28*). Another possible reason is that there is only a small survival chance for people who are caught in fires in the ship. To prevent the occurrence of fires/explosions, ship owners are encouraged to equip their vessels with an efficient flame arrestor, backfire arrester or other similar devices in the carburetors of each inboard engine and efficiently ventilate enclosed spaces so as to remove explosive and flammable gases (*4*). In addition, training is extremely important because there are no fire departments at sea. Crew members and passengers would have to extinguish a fire by themselves.

One interesting finding from this study is that a lower mortality count is associated with contact and machinery/hull damage, shown in Figure 1. Although machinery/hull damage has a small effect in increasing the human life loss in this study, it continues to be the cause of the majority of losses in marine insurance. Statistics from the International Union of Marine Insurance (*29*) reported that 40% of hull claims by number are for machinery damage, accounting for 20% of costs.

Effects of Accident Location. Consistent with the results of the study by Jin, the distance to the shore is also a significant factor influencing the shipping accident consequence with regards to human life loss in this study (*8*). Figure 1 highlights that a sinking accident that occurs far away from the shore can cause 2.14 more fatalities, as compared with a sinking accident close to the shore. This result is consistent with what we expected because there will be a long response time and poor rescuing efficiency for shipping accidents that occur far away from the coastal area/harbor/ports.

Effects of Operation Conditions. Consistent with the results from previous studies, the ship operating time is strongly correlated with human life loss (*16*). Figure 1 shows that the mean of human life loss during the night-time period is generally higher than for the daytime period (by 0.98), which is quite close to the findings from the study conducted by Akten (*1*). The higher human life loss at night might be attributed to the consideration that people require a longer evacuation time at night compared to during the daytime period, especially as most people would be asleep at this time (*30, 31, 32*). Another possible reason for the higher casualty number for nighttime accidents would be that it is more challenging to recover people in the darkness once they are in the water.

Consistent with our expectations, adverse weather conditions are associated with an increase in human life loss. It can be seen from Figure 1 that the expected human life loss is 4.84 higher for sinking accidents that occurred under adverse weather conditions. This implies that for every sinking accident that occurs under adverse weather conditions, 4.84 more fatalities and missing people are expected. Figure 1 also shows a marginal effect of −7.08 for docking conditions, suggesting that there are 7.08 fewer fatalities or missing persons for each sinking accident that occurs when ships are moored or docked.

Conclusions

By analyzing 10 years of worldwide sinking accident data, this study developed a mortality count model to evaluate the number of casualties (i.e., human life loss) resulting from sinking accidents using ZINB regression approaches. The marginal effects of the influencing factors on mortalities were also examined in this study. The mortality count model results show that human life loss is higher for precedent accidents including collision, fire/explosion and capsizing, adverse weather conditions and night-time periods. Furthermore, the marginal effect analysis results show that the increase of the expected human life loss in sinking accidents is the largest for a ship that suffers a precedent accident of capsizing, followed by fire/explosion and collisions. Interestingly, a lower human life loss is associated with contact and machinery/hull damage accidents. Consistent with our expectation, cruise ships involved in sinking accidents usually suffer more human life loss and there will be more human life loss for accidents that occur far away from the coastal area/harbor/port. In addition, fatalities could be reduced when the ship is moored or docked.

One implication of this study is that it is essential to prevent the occurrence of ships capsizing so that the resulting loss can be reduced significantly. Government agencies and shipping companies can employ the developed mortality count model to assess various safety enhancing measures and strategies. The developed model in this study could also help insurance companies to evaluate the possible human life loss and damage to ships. It should be pointed out that the ZINB regression technique is only applicable when there are excess zeros and the distribution of the observed counts is over-dispersed. However, if there is no apparent reason to suspect that two states (i.e., perfect and imperfect states) might be present, the use of a ZINB model may be simply capturing model mis-specification that could result from factors such as unobserved effects (heterogeneity) in the data. One limitation of this study is that shipping traffic information was not included in the collected shipping accident database. In future, we will compare our accident results to traffic properties after collecting shipping traffic data. In the future it would also be interesting to investigate whether the dusk/dawn time period exhibits varying effects on human life loss as compared with the nighttime period.

Acknowledgments

The authors thank the five anonymous reviewers for their thorough review of the earlier versions of the paper. This study is sponsored by the Shuguang Program supported by the Shanghai Education Development Foundation and the Shanghai Municipal Education Commission (Grant No. 16SG41).

Author Contributions

The authors confirm contribution to the paper as follows: study conception and design: T. Chai, J. Weng; data collection: J. Weng; analysis and interpretation of results: T. Chai, J. Weng, X. Chai; draft manuscript preparation: J. Weng. All authors reviewed the results and approved the final version of the manuscript.

References

1. Akten, N. Analysis of shipping casualties in the Bosphorus. *Journal of Navigation*, Vol. 57, 2004, pp. 345–356.
2. Weng, J., and S. Xue. Ship Collision Frequency Estimation in Port Fairways: A Case Study. *Journal of Navigation*, Vol. 68, No. 3, 2015, pp. 602–618
3. Chapman, S. E., and N. Akten. *Marine Casualties in the Turkish Straits- a Way Ahead, Seaways*. The Nautical Institute, London, 1998.
4. Jin, D., H. L. Kite-Powell, and W. K. Talley. The Safety of Commercial Fishing: Determinants of Vessel Total Losses and Injuries. *Journal of Safety Research*, Vol. 32, No. 2, 2001, pp. 209–228.
5. Jin, D., and E. Thunberg. An Analysis of Fishing Vessel Accidents in Fishing Areas Off the Northeastern United States. *Safety Science*, Vol. 43, No. 8, 2005, pp. 523–540.
6. Perez-Labajos, C., B. Blanco, M. Azofra, J. J. Achutegui, and E. Eguia. Injury and Loss Concentration by Sinkings in Fishing Fleets. *Safety Science*, Vol. 47, No. 2, 2009, pp. 277–284.
7. Roberts, S. E., B. Jaremin, and P. B. Marlow. Human and Fishing Vessel Losses in Sea Accidents in the UK Fishing Industry from 1948 to 2008. *International Maritime Health*, Vol. 61, No. 3, 2010, pp. 143–153.
8. Jin, D. The Determinants of Fishing Vessel Accident Severity. *Accident Analysis and Prevention*, Vol. 66, 2014, pp. 1–7.
9. Birpinar, M. E., G. F. Talu, and B. Gonencgil. Environmental Effects of Maritime Traffic on the Istanbul Strait. *Environmental Monitoring and Assessment*, Vol. 152, 2009, pp. 13–23.
10. Aydogdu, Y. V., C. Yurtoren, J. Park, and Y. Park. A Study on Local Traffic Management to Improve Marine Traffic Safety in the Istanbul Strait. *Journal of Navigation*, Vol. 65, No. 1, 2012, pp. 99–112.
11. Weng, J., and D. Yang. Investigation of Shipping Accident Injury Severity and Mortality. *Accident Analysis and Prevention*, Vol. 76, 2015, pp. 92–101
12. Zou, Y., L. Wu, L., and D. Lord. Modeling Over-dispersed Crash Data with a Long Tail: Examining the Accuracy of the Dispersion Parameter in Negative Binomial Models. *Analytic Methods in Accident Research*, Vol. 5–6, 2015, pp. 1–16.
13. Weng, J., E. Y. Ge, and H. Han. Evaluation of Shipping Accident Casualties using Zero-inflated Negative Binomial Regression Technique. *Journal of Navigation*, Vol. 69, No. 2, 2016, pp. 433–448.
14. Anastasopoulos, P. C. Random Parameters Multivariate Tobit and Zero-inflated Count Data Models: Addressing Unobserved and Zero-state Heterogeneity in Accident Injury-severity Rate and Frequency Analysis. *Analytic Methods in Accident Research*, Vol. 11, 2016, pp. 17–32.
15. Eliopoulou, E., and A. Papanikolaou. Casualty Analysis of Large Tankers. *Journal of Marine Science and Technology*, Vol. 12, 2007, pp. 240–250.
16. Talley, W. K., D. Jin, and H. L. Kite-Powell. Determinants of the Severity of Passenger Vessel Accidents. *Maritime Policy and Management*, Vol. 33, 2006, pp. 173–186.
17. Eliopoulou, E., R. Hamann, A. Papanikolaou, and P. Golyshev. Casualty Analysis of Cellular Container Ships. *Proc., IDFS 2013*, Shanghai, 2013, pp. 25–27.
18. Ozsoysal, R., and O. A. Ozsoysal. Maritime Casualties Through the Bosphorus. *Naval Engineers Journal*, Vol. 118, No. 1, 2006, pp. 77–82.
19. Uluscu, O.S., B. Ozbas, T. Altiok, and L. Or. Risk Analysis of the Vessel Traffic in the Strait of Istanbul. *Risk Analysis*, Vol. 29, No. 10, 2009, pp. 1454–1472.

20. Debnath, A. K., and H. C. Chin. Navigational Traffic Conflict Technique: A Proactive Approach to Quantitative Measurement of Collision Risks in Port Waters. *Journal of Navigation*, Vol. 63, No. 1, 2010, pp. 137–152.
21. Weng, J., Q. Meng, and X. Qu. Vessel Collision Frequency Estimation in the Singapore Strait. *Journal of Navigation*, Vol. 65, 2012, pp. 207–221.
22. Lambert, D. Zero-inflated Poisson Regression, with An Application to Defects in Manufacturing. *Technometrics*, Vol. 34, 1992, pp. 1–14.
23. Minami, M., C. E. Lennert-Cody, W. Gao, and M. Roman-Verdesoto. Modeling Shark Bycatch: The Zero-inflated Negative Binomial Regression Model with Smoothing. *Fisheries Research*, Vol. 84, 2007, pp. 210–221.
24. Hall, D. B. Zero-inflated Poisson and Binomial Regression with Random Effects: A Case Study. *Biometrics*, Vol. 56, 2000, pp. 1030–1039.
25. Montewka, J., S. Ehlers, F. Goerlandt, T. Hinz, K. Tabri, and P. Kujala. A Framework for Risk Assessment for Maritime Transportation Systems-A Case Study for Open Sea Collisions Involving RoPax Vessels. *Reliability Engineering & System Safety*, Vol. 124, 2014, pp. 142–157.
26. Lois, P., J. Wang, A. Wall, and T. Ruxton. Formal Safety Assessment of Cruise Ships. *Tourism Management*, Vol. 25, No. 1, 2004, pp. 93–109.
27. House, D. J. *Marine Emergencies: For Masters and Mates*. Routledge, Oxon, 2014.
28. National Research Council. Fishing Vessel Safety: Blueprint for a National Program. National Academy Press, Washington D.C., 1991.
29. Iumi. *Current Issues – Iumi Political Forum*, 2013. http://www.iumi.com/committees/political-forum.
30. Boulougouris, E. K., and A. Papanikolaou. Modeling and simulation of the evacuation process of passenger ships. *Proc., 10th International Congress of the International Maritime Association of the Mediterranean*, Crete, Greece, May 2002.
31. International Maritime Organization. *Guidelines for Evacuation Analysis for New and Existing Passenger Ships*. MSC.1/Circ.1533. London, UK, 2016. http://www.traffgo-ht.com/downloads/pedestrians/downloads/documents/MSC.1,Circ.1533,2016.pdf.
32. Weng, J., G. Li, T. Chai, and D. Yang. Evaluation of Two-Ship Collision Severity using Ordered Probit Approaches. *Journal of Navigation*, Vol. 71, No. 4, 2018, pp. 822–836.

The Standing Committee on Marine Safety and Human Factors (AW040) peer-reviewed this paper (18-06154).

The views expressed in this paper only reflect the opinion of the author and must not be considered as official opinions from any national or international maritime authorities.

Article

TRR
JOURNAL OF THE TRANSPORTATION RESEARCH BOARD

Transportation Research Record
2018, Vol. 2672(11) 73–80

DOI: 10.1177/0361198118794056
journals.sagepub.com/home/trr
SAGE

Determination of Minimum Distance to Obstacle Avoidance in the Singapore Strait

Zhaoyang Lu[1,2], Liujiang Kang[3], Song Gao[4], and Qiang Meng[3]

Abstract

In the maritime simulation models of obstacle avoidance, the parameter of critical contact closest point of approach (CCPA) is defined as the minimum safe distance to avoid an obstacle contact. Although the parameter of CCPA plays a decisive role on the maneuvering of the ship for obstacle avoidance, it is generally set with an uncalibrated input value in many obstacle avoidance simulation models. This study focuses on the parameter calibration of the critical CCPA by utilizing the automatic identification system (AIS) data in the Singapore Strait and maneuver simulation experiments. First, an effective mathematical method is proposed to identify ship obstacle avoidance from the big AIS data. Together with several obstacles and the AIS data in the Singapore Strait, the average critical CCPA values are identified for different ship classes. Second, a series of maneuver simulation experiments on Dalian Maritime University navigational simulators are conducted by professional captains. Based on the experimental data, the simulation-based critical CCPA values are derived. Finally, these calibrated CCPA values from the above two methods are compared and analysed. The comparison result implies that these critical CCPA values, derived from AIS data and maneuver simulators, are matched and confirm the rationality of one another. It also reveals that the average critical CCPA value for each ship class has an increasing trend with the increased ship length.

As one of the most important and busiest shipping waterways in the world, the Straits of Malacca and Singapore (SoMS) are unique and vital to the development of global economics. It is foreseen that more and more large-sized ships will pass through the SoMS. Accordingly, the chance of a navigational accident involving large ships in the SoMS becomes high, and it may lead to catastrophic consequences. To maintain the current navigational safety level, it is crucial to assess the navigation risk when ships pass through the shallow, narrow and busy waterway sectors. In the SoMS, the Singapore Strait is considered as a bottleneck for the shipping routes. When ships pass through the narrow and busy waterway sectors in the Singapore Strait, navigation risk assessment is a high-priority concern for the maritime industries because it is linked to shipping efficiency and port management, as well as environment issues. The navigation risk is also an import metric to evaluate potential shipping routes such as the transarctic shipping routes (*1–4*).

The ship traffic simulation model offers an efficient tool to examine various navigational safety issues. Some ship simulation software packages have been developed as navigation channel design tools. Dand introduced the Permanent International Association of Navigation Congress (PIANC) simulation approach for water channel design (*5*). Köse et al. developed a discrete event simulation based model for ship traffic in the Istanbul Strait (*6*). Ince and Topuz proposed a safe navigation design model incorporating vessel arrivals and the hydrodynamic force mechanisms (*7*). Considering that these models neglected the ship interactions and operation behaviors of captains, Szpak and Tapamo described the tracking route of moving ships in water areas based on a proposed background-subtraction model (*8*). Qu and Meng used a discrete choice model to estimate the loss to the global economy on the hypothesis of an extreme scenario of blockade of the SoMS, they then developed a

[1]School of Economics and Management, Southwest Jiaotong University, Chengdu, PR China
[2]Centre for Maritime Studies, National University of Singapore, Singapore
[3]Department of Civil and Environmental Engineering, National University of Singapore, Singapore
[4]Intelligent Transportation Systems Research Center, Wuhan University of Technology, Wuhan, China

Corresponding Author:
Address correspondence to Qiang Meng: ceemq@nus.edu.sg

ship traffic simulation model to capture the behaviors of captains and ship movement in the Straits (*9*, *10*).

However, the majority of these simple models are created based on their assumptions without any calibration and validation. To some extent, many of the proposed methods for identification and quantification of sensitive situations are too subjective. This may be attributed to the fact that it is impossible to have access to all of the data that could potentially determine the likelihood of an accident or its resulting injury severity (*11*). On the other hand, it is well known that the automatic identification system (AIS) data is a significantly important source of information for addressing maritime traffic problems and associated ship collision situations. Felski and Jaskolski showed the effectiveness of AIS data in ship collision avoidance, and presented a discussion concerning the incompleteness and integrity of AIS data (*12*). Simulation models and algorithms for collision avoidance will be more convincing if they are based on empirical data (*13*). Based on the navigation simulator-based experiments and empirical data, we can extract more accurate and reasonable operation rules of real ship movements. To improve the simulation accuracy, addressing the calibration of input parameters of these simulation models is a meaningful and urgent problem, but few studies have made use of AIS data for parameter determination of the navigation simulation models.

This study focuses on parameter determination of the critical contact closest point of approach (CCPA), namely, the minimum distance to obstacle avoidance, by utilizing the AIS data in the Singapore Strait and maneuver simulation experiments. In practice, this parameter of CCPA may change with the ship size, type, speed, maneuverability, sea state, and weather conditions. Different navigators may take a different CCPA to avoid collisions. Note that, if these CCPAs are smaller than the critical CCPA, the travelling vessels will have a high risk of obstacle contact. As a result, the determination of the critical CCPA will have a direct effect in making the simulated obstacle contact behaviors more reliable and similar to actual behaviors.

Definition of Critical CCPA

The parameter of critical CCPA is defined as the practical minimum distance that vessels are able to safely avoid an obstacle contact. That is, the type of closest point of approach (CPA) is specific to obstacle avoidance, and the critical CCPA may be different for different vessel classes. In other words, the CCPA of a different vessel class cannot be lower than their critical CCPA for navigation safety, either in the ship traffic navigation or simulation models. In fact, researchers have proposed various methods to simulate an obstacle avoidance shipping route, such as the potential filed method (*14–16*), or the neurofuzzy networks predictor method (*17*). In these obstacle avoidance studies, although the CCPA plays a decisive role on the avoidance safe route, it is set with an uncalibrated input value in general. Thus, the determination of the critical CCPA is required before using these simulation models.

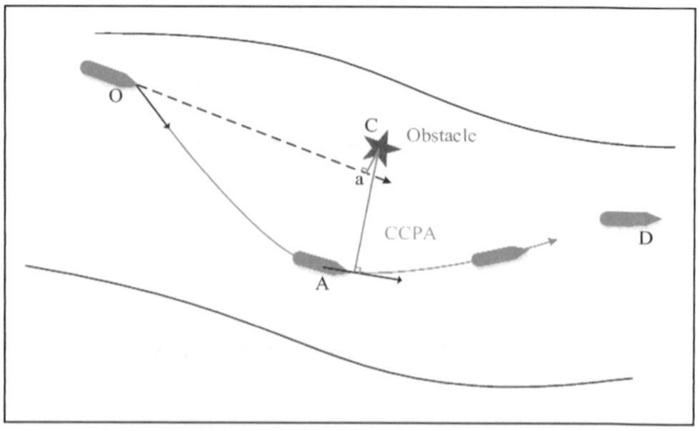

Figure 1. Illustration of obstacle avoidance and CCPA.

To be more intuitive, let us consider an obstacle in a direct shipping route between the origin point and destination point of one ship. As Figure 1 shows, for the original route, the minimum distance to the obstacle is the solid line *Ca*, which is regarded as unsafe by the captain. Therefore, the captain needs to change this shipping route, and the minimum distance to the obstacle becomes the orange line *CA*. Here, the line of *CA*, whose distance could just allow the ship to safely avoid the obstacle contact, is defined as the critical CCPA. Accordingly, it can be seen that the critical CCPA directly determines the shortest actual shipping route, shown in the full blue line. To sum up, this study proposes the following two methods for the derivation of critical CCPA from the empirical and simulation aspects, respectively:

(a) An effective method is proposed to identify ship obstacle avoidance from the big AIS data. Then, based on several obstacles in the Singapore Strait, the critical CCPA values are identified for different ship classes.

(b) A series of maneuver simulation navigational experiments on Dalian Maritime University (DMU) navigation simulators are conducted by eight professional captains. Note that these captains know that the aim of these experiments is to obtain the critical CCPA.

Based on the practical AIS data, the CCPA of each vessel can be identified by making use of our proposed method. On the other hand, in the DMU navigation simulation

Table 1. Information of Fixed Obstacles in the Singapore Strait

ID	Name	Position	Longitude	Latitude	Study region
1	Takong beacon	Leg 9E	103.728	1.097	(103.728 ± 0.01, 1.097 ± 0.01)
2	Racon LF1.15	Leg 9E	103.648	1.060	(103.648 ± 0.01, 1.060 ± 0.01)
3	Helen mar reef racon	Leg 11E	103.775	1.124	(103.775 ± 0.01, 1.124 ± 0.01)
4	Buffalo rock	Leg 11E	103.802	1.166	(103.802 ± 0.01, 1.166 ± 0.01)
5	Batu berhanti beacon	Leg 12E	103.875	1.196	(103.875 ± 0.01, 1.196 ± 0.01)
6	Horsburgh lighthouse	Leg 15E	104.405	1.330	(104.405 ± 0.01, 1.330 ± 0.01)
7	Raffile lighthouse	Leg 5W	103.741	1.160	(103.741 ± 0.01, 1.160 ± 0.01)
8	Tanjung piai beacon	Leg 7W	103.510	1.259	(103.510 ± 0.01, 1.259 ± 0.01)

experiments, the captains will try to choose the minimum distance to avoid an obstacle, as they know that is the aim of these experiments. Then, if the derived critical CCPAs from AIS-data are matched with our results from the DMU navigation simulators, they will confirm the rationality of each other.

AIS Data Based CCPA Derivation

The AIS data used in this study is composed of the ship movement data collected in the Singapore Strait, as well as the ship information. This data set has been used to carry out several empirical studies on the SoMS (*9*, *18–20*). To be more specific, the movement data includes Maritime Mobile Service Identity (MMSI), ship position, course, draught, recorded time and speed. The ship information data contains the characteristics of 70,509 ships, including their MMSI number, length, beam, type of ship, and so forth. In short, it is sufficient to use these data to derive the critical CCPA. Based on the AIS data of the Singapore Strait from October 1, 2013, to October 7, 2013, this section will introduce the method used for identifying ship obstacle avoidance data and the derived CCPA values.

Method of Identifying Obstacle Avoidance CCPA

In this section, a five-step method is proposed to identify the critical values of obstacle CCPA for different ship classes, as is shown in Figure 2.

Now, we will elaborate each of the five steps. Step 1 defines the study region of the obstacle, based on its longitude and latitude position. First of all, several stationary obstacles in the Singapore Strait, such as radar beacon, lighthouse, and lighted beacon, are noted with their location information, as illustrated in Table 1. In this table, the obstacle position information refers to the 15 legs of the Singapore Strait, which are identified by the Society

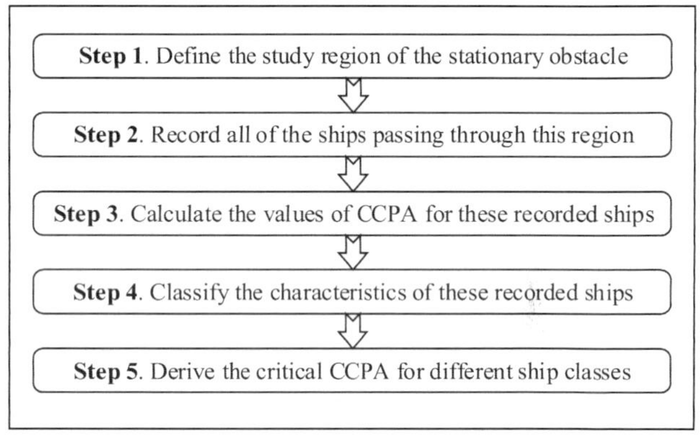

Figure 2. Flowchart of identifying the critical obstacle CCPA.

of International Gas Tanker and Terminal Operators Ltd (*21*). Then, a study region in the waterway can be defined based on the obstacle location. As Figure 3 shows, the obstacle position is denoted by (X,Y), and the region step is noted as (sx, sy) for the longitude and latitude, as a result, the study region is (X ± sx, Y ± sy). In this experiment, (sx, sy) is set as (0.01°, 0.01°), a region of about 2226 m × 2000 m.

Next, Step 2 records all of the ships passing through the study region during a certain time period per day (e.g. 8:00 AM–11:00 AM) in the AIS data, and Step 3 calculates the values of CCPA for these recorded ships. That is, when the recorded position of a ship is located in the study region, we will record its MMSI. Then, we will calculate the distance between the ship and the obstacle. Here, it is worth mentioning about the method of calculating the distance between point C and point A. By assuming the longitudes and latitudes of Points C and A as (C_x, C_y), (A_x, A_y), the distance between the two points can be derived in the unit of meters, by making use of the following ellipsoidal-surface equations (*22*, *23*):

$$\begin{cases} \text{Sin } X = \text{Sin}(A_y\ \pi/180) \times \text{Sin}(C_y\ \pi/180) \\ \text{Cos } X = \text{Cos}(A_y \times \pi/180) \times \text{Cos}(C_y \times \pi/180) \times \text{Cos}(C_x \times \pi/180 - A_x \times \pi/180) \end{cases} \quad (1)$$

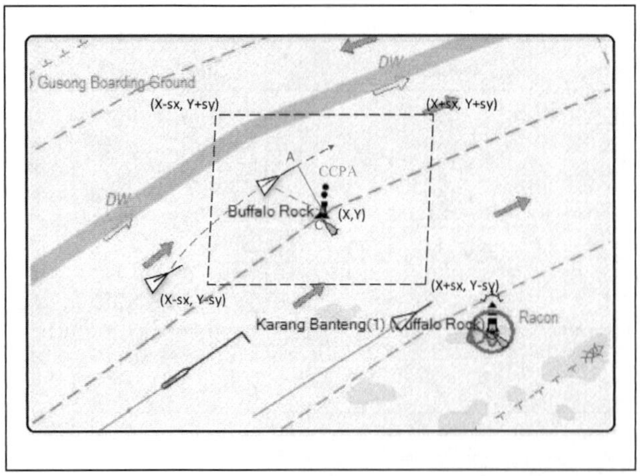

Figure 3. Illustration of how to identify obstacle avoidance data.

$$Dis = 6368.16 \times 10^3 \times \left[ArcTan\left(\frac{-(\mathrm{Sin}\ X + \mathrm{Cos}\ X)}{\sqrt{1 - (\mathrm{Sin}\ X + \mathrm{Cos}\ X)^2}} \right) + 2 \times Arc\ \mathrm{Tan}(1) \right] \quad (2)$$

As the position data of ships is renewed at least every 30 seconds, the shortest distance between the ship and the obstacles can be seen as the value of CCPA. We now have the CCPA of all ships in the study region, but this may not be the critical CCPA.

Finally, Step 4 classifies the recorded ships into different classes based on their length characteristics, and Step 5 identifies the average critical CCPA for different ship classes. In this study, the ships are classified into six classes according their length. Following the ship length based "Filter" classification on the "*Shipfinder*" website, the six ship length intervals are 0–40 m, 41–80 m, 81–160 m, 161–240 m, 241–320 m, and over 320 m (*24*). Then, from the minimum CCPA of each ship class during our everyday selected period of one week, the average value of the critical obstacle CCPA is obtained for each ship class.

Results of the Obtained Values of CCPA

Based on the above five-step introduced method and the AIS data, we can identify various contact avoidance shipping routes and their critical obstacle CCPA values. For example, Figure 4 illustrates several contact avoidance shipping routes near the Batu Berhanti Beacon in the Leg12E of the Singapore Strait. By using our proposed method, these routes are derived from different ships recorded in the AIS data. Then, the average critical CCPA for each ship class is calculated by including all of the obstacles listed in Table 1.

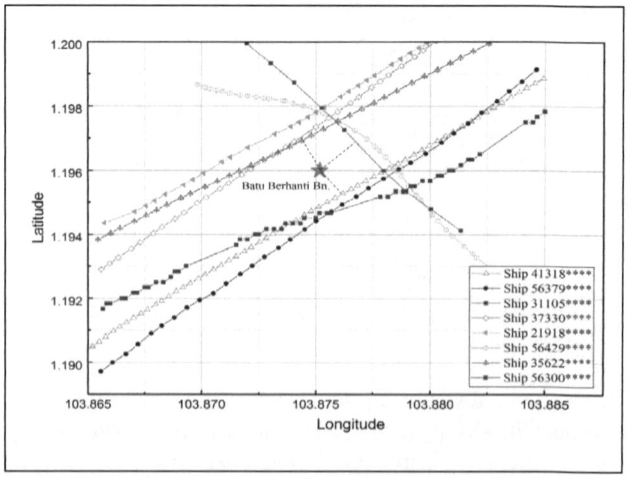

Figure 4. Examples of AIS data recorded contact avoidance shipping routes near the obstacle of Leg12E (Collection Period: October 1–7, 2013).

In summary, Table 2 tabulates the AIS-data based critical CCPA values for all of the ship classes in the Singapore Strait, which contains the collected period, the mean and variances of CCPA values, and main collected positions. Firstly, based on the AIS data, the mean and variances of the critical CCPA values are derived for each ship class. Note that as these values are data-based, they may have some variations in a different database. Secondly, the average speeds of most ships are close to 12 knots, which is the maximum speed limitation in this strait. The exceptional case is the small boats of Class 1, which navigate with a greater speed (17 knots), even when they are near the obstacle. Thirdly, the collection positions reveal that when a fixed obstacle is located near the center position, not the two sides of a waterway, it will be easier to obtain valid data points. For example, the obstacles in Leg12E, Leg 9E, and Leg 11E are all near the center of this waterway.

Maneuver Simulation Experiments

To further evaluate the values of critical CCPA, we conducted several navigational experiments in the industry key laboratory of traffic and transportation on the DMU navigation simulator. All of these maneuver simulation experiments were navigated by eight professional captains, who have more than 10-year navigating experience, and know the experiment objectives. Four scenes of the DMU simulators are illustrated in Figure 5. Regarding the physical environment setting, it mainly included the simulated scene, wind, flow, fog and wave conditions, and experimental time. In our experiments, the simulation scene was set as the Singapore Strait, and the experiment time was January 16–18, 2017. It should be noted

Table 2. AIS-Data Based Critical Obstacle CCPA in the Singapore Strait (Collection Period: October 1–7, 2013)

Class	Length (m)	Aver speed (knot)	Mean CCPA (m)	Variance CCPA (m^2)	Collected position
1	0–40	17.00	193.56	85.76^2	Leg12E, 9E, 13E, 15E
2	40–80	11.81	214.43	92.05^2	Leg 12E, 9E, 13E
3	80–160	11.65	218.85	64.04^2	Leg 12E, 13E
4	160–240	11.20	223.67	91.63^2	Leg 12E, 9E, 13E, 5W
5	240–320	12.25	242.98	77.20^2	Leg 12E, 11E, 9E, 13E
6	>320	11.23	271.78	67.45^2	Leg 9E, 11E, 13E

Figure 5. Navigation simulators in the DMU.

that the following parameters referred to the weather conditions of October 1–7, 2013, in which the wind direct is set from South to North, 1.5 km/h, the flow direction is set as 180°, 1.5 km/h, no fog, and the wave condition is set as calm-glassy.

In our designed navigation simulator experiments, with the predetermined settings, the captains manipulated the simulators to avoid obstacle contact. Then, the simulator recorded the AIS data of their travelling route, including the ship position, course, draught, recorded time, speed and so on. Similar to the calculation algorithm of AIS-data, the critical CCPA can be derived from these simulation experiments. For the reason that there are sufficient small ships in the collected AIS data, our manoeuvre simulation experiments mainly focused on the large ships in Class 5 and 6. The input parameters for the navigational experiments, such as ship size, course, class, and speed, are tabulated in Table 3. The simulation results of these six experiments recorded the captains' actions to help ships avoid obstacle contact. In this study, as Table 3 shows, these action parameters mainly include the travel course change, time to the closet point of approach (TCPA), and the values of CCPA. In short, the six experiments reported that the critical CCPA values for ships of Class 6 are from 222.36 m to 240.89 m, for ships of Class 5 are from 185 m to 259 m, and for ships of

Table 3. DMU Maneuver Simulation Based Critical Obstacle CCPA

Ship ID	Size (m)	Class	Course	Speed (knot)	Course change	TCPA (min)	CCPA (m)
CO1	~340 m	6	258°	12	−9.1°	6.89	222.36
CO2	~336 m	6	258°	12	−3.7°	9.26	240.89
CO3	~300 m	5	258°	12	6.1°	18.28	185.3
CO4	~280 m	5	258°	12	−9.0°	4.09	240.89
CO5	~292 m	5	258°	12	7.2°	19.42	259.42
CO6	~120 m	3	258°	12	4.6°	5.72	166.77

Table 4. Differences of the Critical CCPA Values Obtained from the AIS-Data and DMU Simulators

		Critical CCPA (m)				
Ship ID	Class	DMU simulated	AIS mean	AIS Standard variance	AIS-DMU differences	Result match or not
CO1	6	222.36	271.78	64.04	49.42	Yes
CO2	6	240.89	271.78	64.04	30.89	Yes
CO3	5	185.3	242.98	77.2	57.68	Yes
CO4	5	240.89	242.98	77.2	2.09	Yes
CO5	5	259.42	242.98	77.2	−16.44	Yes
CO6	3	166.77	218.85	67.4	52.08	Yes

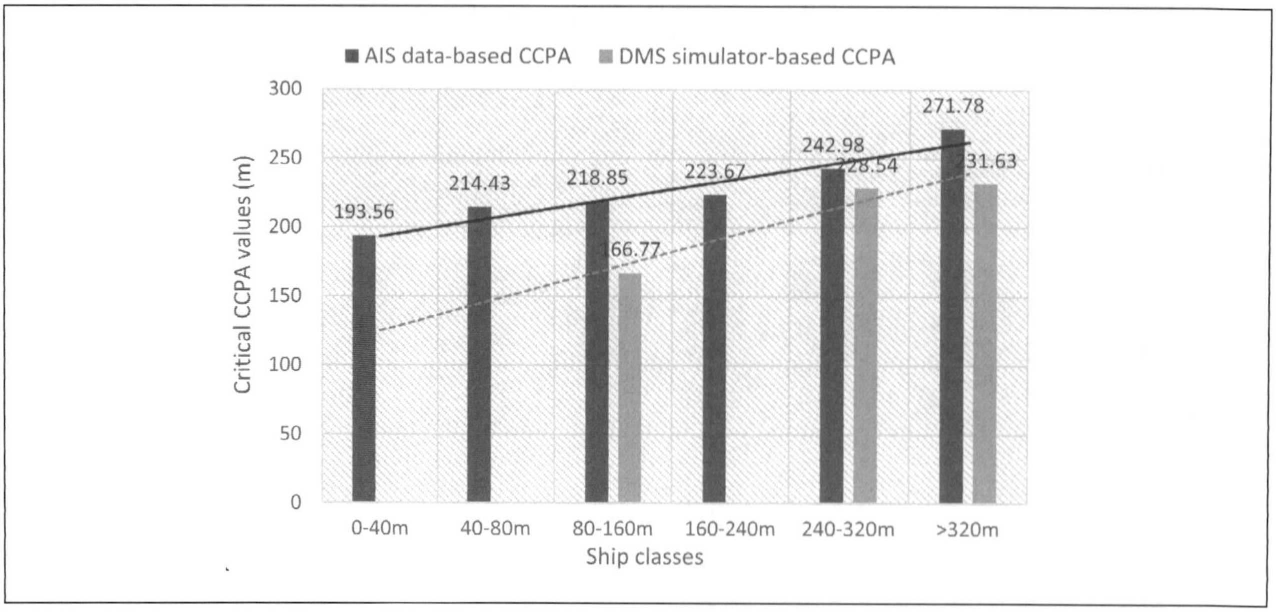

Figure 6. Column chart of critical obstacle CCPA values for different ship classes.

Class 3 they are 166.77 m. The variances of the critical CCPA value may be attributed to the different operational experiences and judgment of different captains.

Result Comparisons and Discussions

Previously, the values for the critical obstacle CCPA have been derived from both the AIS data and the DMU manoeuvre simulators. Table 4 lists the comparison of these values. It shows that, for all of the six simulated ships, the differences between the simulation-based and AIS-data based critical CCPA are less than the standard variances calculated from the AIS-data. In other words, all of these simulation values may appear in the historical data. Thus, it is believed that the results of critical CCPA values obtained from the AIS-data and DMU simulators are matched and could confirm each other's rationality.

Figure 6 shows the average critical CCPA values for all the six length-based ship classes. It can be found that with the increase of the ship length, the average critical CCPA values reveal an increasing trend. Obviously, the mean critical CCPA values are increased from 193.56 m for Class 1 to 271.78 m for Class 6. This characteristic is similar to the character of the ship's domain. As Fujii and Tanaka pointed out, the ship domain has an elliptical shape, whose major axis and minor axis are a linear function of the ship length (*25*). To some extent, the distance of CCPA is another kind of obstacle related ship domain. In fact, with the increase of the ship length, the minimum distance needed to safely avoid obstacles is increased. This also explains the reason why ships are classified according to their lengths in this study.

In addition, Figure 6 also illustrates the average critical CCPA values for the three simulated ship classes, that is, Class 3, 5 and 6. One remarkable feature is that these values keep increasing with the increase of the ship length, which is consistent with the obtained results from AIS data. Another distinctive feature is that for each ship class, the mean AIS-data based critical CCPA is a little larger than the DMU simulated value in general. This may be because, during the maneuver simulation, the captains have to know the simulation objectives so they can try to choose the minimum distance to avoid obstacles. However, in practice, most captains would choose a larger distance to avoid obstacle contact. Fortunately, these differences for each ship class are less than the standard variance obtained from AIS data, so they are within the acceptable range.

Conclusions

This study discussed determination of the critical CCPA by making use of the AIS data in the Singapore Strait and maneuver simulation experiments. In general, this parameter was derived from the following two aspects: (a) based on the collected AIS data in the Singapore Strait, an effective five-step method was proposed to identify the critical CCPA values for different ship classes; (b) several maneuver simulation experiments in DMU navigation simulator were conducted by eight professional captains to obtain the critical CCPA values. Finally, the comparison results of the derived critical CCPAs from AIS data and maneuver simulators revealed, to some extent, the rationality of each method. It was also found that the obtained average critical CCPA value for each ship class increases with the growth of ship length.

The determination of the critical average CCPA for different length-based ship classes is a meaningful research issue with practical significance. The contributions of this study can be further expressed as follows. Firstly, based on our proposed method and collected data in the Singapore Strait, the critical CCPA values were derived from both AIS-data and maneuver simulators, and the comparison results confirmed the rationality of each method. Secondly, the values of CCPA have a direct effect on the obstacle avoidance safe routes. Our derived critical CCPA values could provide a reliable reference for the input values used in the automatic navigation simulation models.

The obstacle contact is only one of the common navigation collision situations. This study calibrated the critical CCPA in a simplified condition, by neglecting some unexpected external factors, such as weather conditions, visibility, time period, and source information. It would be worthwhile to further investigate this topic by accounting for the varieties of these external conditions along an internationally accepted and defined spectra. Moreover, the consequence of ship accidents also includes ship collision (e.g. crossing collision, heading-on collision, and overtaking collision), and ship grounding. We will continue to investigate ship collision avoidance rules and parameters calibration, based on the big AIS data in the Singapore Strait and DMU maneuver simulators. Further research is needed to develop some effective methods to identify ship interaction rules from the big AIS database.

Acknowledgments

This study is supported by the research project "Impact Analysis of Large Ships on Ship Traffic in the Straits of Malacca and Singapore" (Project ID: SMI-2015-MA-03) funded by the Singapore Maritime Institute.

Author Contributions

The authors (Zhaoyang Lu, Liujiang Kang, Song Gao and Qiang Meng) confirm contribution to the paper as follows: study conception and design: Zhaoyang Lu, Lijiang Kang, Song Gao, Qiang Meng; data collection: Zhaoyang Lu, Lijiang Kang, Song Gao, Qiang Meng; analysis and interpretation of results: Zhaoyang Lu, Liujiang Kang, Song Gao and Qiang Meng; draft manuscript preparation: Zhaoyang Lu, Liujiang Kang, Song Gao and Qiang Meng. All authors reviewed the results and approved the final version of the manuscript.

References

1. Zhang, Y., Q. Meng, and S. H. Ng. Shipping Efficiency Comparison between Northern Sea Route and the Conventional Asia-Europe Shipping Route via Suez Canal. *Journal of Transport Geography*, Vol. 57, 2016, pp. 241–249.
2. Zhang, Y., Q. Meng, and L. Zhang. Is the Northern Sea Route Attractive to Shipping Companies? Some Insights from Recent Ship Traffic Data. *Marine Policy*, Vol. 73, 2016, pp. 53–60.

3. Meng, Q., Y. Zhang, and M. Xu. Viability of Transarctic Shipping Routes: A Literature Review from the Navigational and Commercial Perspectives. *Maritime Policy & Management*, Vol. 44, No. 1, 2017, pp. 16–41.
4. Wang, H., Y. Zhang, and Q. Meng. How Will the Opening of the Northern Sea Route Influence the Suez Canal Route? An Empirical Analysis with Discrete Choice Models. *Transportation Research Part A: Policy and Practice*, Vol. 107, 2018, pp. 75–89.
5. Dand, I. Approach Channel Design: The PIANC Approach. *Proc., International Workshop on Channel Design and Vessel Manoeuvrability*, Norfolk, VA., 2001.
6. Köse, E., E. Başar, E. Demirci, A. Güneroğlu, and Ş. Erkebay. Simulation of Marine Traffic in Istanbul Strait. *Simulation Modelling Practice and Theory*, Vol. 11, No. 7, 2003, pp. 597–608.
7. Ince, A., and E. Topuz. Modelling and Simulation for Safe and Efficient Navigation in Narrow Waterways. *The Journal of Navigation*, Vol. 57, No. 1, 2004, pp. 53–71.
8. Szpak, Z. L., and J. R. Tapamo. Maritime Surveillance: Tracking Ships inside a Dynamic Background Using a Fast Level-Set. *Expert Systems with Applications*, Vol. 38, No. 6, 2011, pp. 6669–6680.
9. Qu, X., and Q. Meng. The Economic Importance of the Straits of Malacca and Singapore: An Extreme-Scenario Analysis. *Transportation Research Part E: Logistics and Transportation Review*, Vol. 48, No. 1, 2012, pp. 258–265.
10. Qu, X., and Q. Meng. Development and Applications of a Simulation Model for Vessels in the Singapore Straits. *Expert Systems with Applications*, Vol. 39, No. 9, 2012, pp. 8430–8438.
11. Mannering, F. L., V. Shankar, and C. R. Bhat. Unobserved Heterogeneity and the Statistical Analysis of Highway Accident Data. *Analytic Methods in Accident Research*, Vol. 11, 2016, pp. 1–16.
12. Felski, A., and K. Jaskolski. The Integrity of Information Received by Means of Ais During Anti-Collision Manoeuvring. *TransNav: International Journal on Marine Navigation and Safety of Sea Transportation*, Vol. 7, No. 1, 2013, pp. 95–100.
13. Nishimoto, T., K. Mukaigawa, S. Tominaga, N. Lubbe, T. Kiuchi, T. Motomura, and H. Matsumoto. Serious Injury Prediction Algorithm Based on Large-Scale Data and under-Triage Control. *Accident Analysis & Prevention*, Vol. 98, 2017, pp. 266–276.
14. Shi, C., M. Zhang, and J. Peng. Harmonic Potential Field Method for Autonomous Ship Navigation. *Proc., 7th International Conference on ITS Telecommunications*, IEEE, Sophia Antipolis, France, 2007, pp. 1–6.
15. Lee, S.-M., K.-Y. Kwon, and J. Joh. A Fuzzy Logic for Autonomous Navigation of Marine Vehicles Satisfying Colreg Guidelines. *International Journal of Control Automation and Systems*, Vol. 2, 2004, pp. 171–181.
16. Xue, Y., D. Clelland, B. S. Lee, and D. Han. Automatic Simulation of Ship Navigation. *Ocean Engineering*, Vol. 38, No. 17–18, 2011, pp. 2290–2305.
17. Harris, C. J., X. Hong, and P. A. Wilson. An Intelligent Guidance and Control System for Ship Obstacle Avoidance. Proc Imeche Part I. *Journal of Systems and Control Engineering*, Vol. 213, No. 14, 1999, pp. 311–320.
18. Zhang, L., Q. Meng, and T. F. Fwa. Big Ais Data Based Spatial-Temporal Analyses of Ship Traffic in Singapore Port Waters. *Transportation Research Part E: Logistics and Transportation Review*, 2017, in press.
19. Kang, L., Q. Meng, and Q. Liu. Fundamental Diagram of Ship Traffic in the Singapore Strait. *Ocean Engineering*, Vol. 147, 2018, pp. 340–354.
20. Qu, X., Q. Meng, and L. Suyi. Ship Collision Risk Assessment for the Singapore Strait. *Accident Analysis & Prevention*, Vol. 43, No. 6, 2011, pp. 2030–2036.
21. SIGTTO. *Passage Planning Guide – Straits of Malacca and Singapore (Soms) (Ppg-Soms 2015 Edition)*. Witherby Seamanship International Ltd., Scotland, UK, 2015.
22. Rapp, R. H. *Geometric Geodesy Part I*. Ohio State University Department of Geodetic Science and Surveying, Columbus, OH, 1991.
23. Lambert, W. D. The Distance between Two Widely Separated Points on the Surface of the Earth. *Journal of the Washington Academy of Sciences*, Vol. 32, No. 5, 1942, pp. 125–130.
24. Shipfinder. Http://Www.Shipfinder.Com/. Accessed June 07, 2017.
25. Fujii, Y., and K. Tanaka. Traffic Capacity. *The Journal of Navigation*, Vol. 24, No. 4, 1971, pp. 543–552.

The Standing Committee on Marine Safety and Human Factors (AW040) peer-reviewed this paper (18-03409).